Liquid Soapmaking

Liquid Soapmaking

Tips, Techniques and Recipes for Creating All Manner of Liquid and Soft Soap Naturally!

Jackie Thompson

Goldilocks Press

Copyright © 2014 Jackie Thompson

www.liquidsoapmakingbook.com

Goldilocks Press
P.O. Box 426
Talco, TX 75487

All rights reserved. No part of this publication may be reproduced, distributed, or transmitted in any form or by any means, including photocopying, recording, or other electronic or mechanical methods, without the prior written permission of the publisher, except in the case of brief quotations embodied in critical reviews and certain other noncommercial uses permitted by copyright law.

ISBN: 978-0-9903115-0-8

1st printing 2014
2nd printing 2015

Cover and Book design by RD Studio

Editor: Kerri Mixon

Photography: Gaily Baughman, Jami Birdsong and Jackie Thompson

---

While the author and publisher have made every effort to ensure the information in this book is correct and free of any error or omission, the information is provided without warranty either expressed or implied, including the implied warranties of merchantability and fitness for a particular purpose.

As described in "Chapter 1: Safety First," some of the chemicals and compounds referred to in this book are dangerous substances which can cause injury if not used with appropriate safety equipment and in accordance with the manufacturer's directions and warnings. The author and the publisher shall not be liable, directly or indirectly, to any person for any direct, incidental or consequential loss, damage, injury or violation of law resulting from the use of the information, directions, or chemicals and compounds contained in this book, and liability where otherwise obtaining, and in all events whatsoever, shall be limited to the purchase price of this book.

Any links to websites are provided for informational purposes only and do not constitute the endorsement of any products or services provided by these websites. The links are subject to change and may expire or be redirected without notice.

In Memory of My Father
JACK DAVIS

"If we stand tall, it is because we stand on the shoulders of those who come before us"

∼ OLD YORUBA PROVERB

CONTENTS

Foreword ix
Preface xi
Acknowledgments xiii

1 Safety First 1
2 Introduction to Potassium Soaps 7
3 How to Make Liquid Soap Using the Paste Method 23
4 Soap Paste Recipes 45
5 How to Make Liquid Soap Using the No Paste Method 59
6 No Paste Recipes 65
7 Thickening Liquid Soap 73
8 How to Make Soap Gels and Jellies 79
9 Soap Gels and Jellies Recipes 85
10 Coloring Liquid Soap 95
11 Fragrancing Soap Naturally 99
12 Incorporating Additives 139
13 Shelf Life and Stability 149
14 Specialty Soap 165
15 How to Formulate Liquid Soap Recipes 177
16 Troubleshooting Problems 187

Appendices 191
Glossary 197
Bibliography 205
Resources 209
Index 211

FOREWORD

As a professional soapmaker who has made countless gallons of liquid soap by hand, it is my opinion Jackie Thompson's book will soon become an industry standard. While reading this book, I was aware of being excited at the new horizons Jackie Thompson has presented for me to discover. Her intelligible and descriptive chapters on "no paste" liquid soapmaking and gel soapmaking kept me enthralled.

We soapmakers now benefit from Jackie Thompson's generosity and willingness to share the end result of her laborious experiments, such as the ingenious introduction of potassium carbonate as an accelerant to directly make a liquid soap without first making a paste. The "no paste" chapters and recipes are all-inclusive and contain valuable information on adding ingredients to speed the reaction, eliminating the need for mechanical mixing and saving the soapmaker valuable time.

Gel soap, the product that has previously eluded many soapmakers, is presented in understandable steps, reliably resulting in the consistent creation of gel soap. The text is so thoroughly complete and comprehensible, even new soapmakers will confidently embark on the artful journey of making gel soap with sureness.

In addition to the new prospects of "no paste" and gel soap, Jackie Thompson's book includes other valuable information pertinent to handmade liquid soap, such as recipes for household cleansers, a section covering additives to liquid soap, scenting naturally with essential oils, and options for different preservatives and their respective effectiveness. Soapmakers of all skill levels will benefit from this fabulously extensive book, which should soon become the "bible" for making liquid soap.

—Kerri Mixon
Soapmaker, Entrepreneur, Author and Editor

PREFACE

Looking back on the past three years spent studying the vagaries and metamorphosis of liquid soap, I can only think how full of myself I was at the beginning of this journey. Research requires a disciplined methodical mind, not a pinball ricochet of ideas exploding at 2 am and half-forgotten before that first swallow of caffeine. Yet somehow, some way, the methods evolved—most often while looking for something else and frequently by total accident. Very few trial batches failed to teach me something new or exciting about this fascinating field, but never was one question answered without giving rise to another. The possibilities are almost endless and it is difficult to find a reasonable stopping place. I leave it to my readers to carry forward.

ACKNOWLEDGMENTS

Special thanks to:

- Catherine Failor, for unlocking and opening the door to liquid soapmaking for the handcrafted soapmaking community. It is a precious gift you gave!

- My husband Stan, for learning more than he ever wanted to know about liquid soap.

- My family and friends for their support and encouragement.

- The Handcrafted Soap and Cosmetic Guild for all the wonderful and valuable information they have taught me along the way.

- To all my soapmaking friends who have answered questions and given me tips and techniques that helped in the writing of this book. You are many and you are mighty!

- Michael at Design for Books. You have given my book character and beauty through your brilliant design. Thanks for going above and beyond your duties as a graphic designer with your sound advice and direction.

- Marie Gale and Kevin Dunn for taking the time to read over the book and give me their invaluable suggestions and insights! Any errors in the interpretation of their advice are my own and shouldn't reflect upon their reputations.

- Kerri Mixon for pulling it all together with her skillful editing. As a seasoned soapmaker she gave my book the clarity and polish it so desperately needed!

CHAPTER

SAFETY FIRST

*"Soap may be fairly harmless and mild, but
its components go through some very active,
caustic stages before they are tamed."*

~ *The Soapmaker's Companion* (CAVITCH, 1997)

Being safe and making safe products is the number one priority when making handcrafted soap. Whether you are a beginner or veteran soapmaker, taking the time to read and follow the recommendations on the next few pages will help assure a trouble-free soapmaking experience.

Setting up Shop

Whether your shop is located in your kitchen, garage or an elaborate manufacturing facility, the workspace should be clean and orderly. Following good manufacturing practices will protect your products from contamination and help prevent accidents brought about by chaos.

Your work area should be located in an area with good ventilation and away from children, pets and other distractions that prevent you from focusing on the task at hand.

Electrical outlets should be conveniently located and sufficient in number for the necessary appliances. Extension cords should be avoided whenever possible, but when necessary they should be appropriately secured to prevent the cord from being dragged or pulled across the work area. To avoid tripping, never place an extension cord across walkways or aisles.

The work area should be equipped with a safety shower or eye wash station in the event of a chemical spill.

A fire extinguisher designed for the use of solvent and grease fires should be placed or mounted in a highly visible area three and one half to five feet above the floor and away from small children. The fire extinguisher should not be too close to a potential source of fire as the fire may prevent the soapmaker from reaching the extinguisher.

Keep floors clean and dry to avoid slips and falls. Clean up spills as they occur and provide matting and/or floor protection in areas prone to grease and oil spills. Keep pathways clear of supplies and equipment to avoid tripping.

All supplies should be stored in appropriate containers off of the floor. They should be clearly marked for your own use and safety as well as the safety of any First Responders in the event of a fire or emergency. A Material Safety Data Sheet (MSDS) should be on file and accessible for all ingredients. Potentially harmful ingredients should be secured safely away from children and pets.

All utensils and containers used for soapmaking should only be used for soapmaking. They should be stored separately from utensils and containers used for preparing or serving food and beverages.

Educate children and all family members about the ingredients being used. Make sure they are aware many of the ingredients could be very dangerous, even fatal if not handled properly.

Personal Safety Equipment

Caustic soda beads can bounce as high as the top of your head when spilled on a countertop! Accidents happen in the blink of an eye even to the most experienced soapmaker. Wearing the appropriate safety gear at all times can mean the difference between a tragedy and an inconvenience.

Safety Goggles: It's important to note safety goggles are not the same as safety glasses. Safety glasses just cover the eyes. Safety goggles enclose the eye area with a protective plastic or rubber ring and are better at protecting from splashes (and bouncing caustic soda beads). Safety goggles are the main defense against eye injury. Some things are irreplaceable; vision is one of them. Always wear safety goggles!

Gloves: The feeling of raw soap on unprotected hands can easily be compared to sticking your hand in a fire ant hill! If that is not enough of a deterrent, consider all of the chemicals you come in contact with as a soapmaker and what effect it will have if transferred to another's more delicate skin or onto the food you prepare. Wear gloves to protect yourself as well as those around you.

Mask and Ventilation: When potassium or sodium hydroxide is mixed with water, the vapor from the exothermic reaction contains particles of alkali that irritate and burn the lungs. Continued exposure can cause permanent damage. Reducing the temperature by using ice in the place of water or adding the hydroxides more slowly may eliminate most of the fumes, but always mixing the lye solution in a well-ventilated area and wearing a mask to protect the lungs from the dust of dry potassium or sodium hydroxide is a better policy. Wearing a chemical respirator rated to protect against alkali fumes is the best policy.

Clothing: Your clothes should protect you just as your gloves, masks and goggles protect you. Loose and flowing shirts and jackets have no place around a soap pot, for they will invariably find a way to get into the soap pot. The more skin is exposed, the more likely it will eventually get burned. Sleeves should be long and fit tightly enough that the gloves easily fit over the sleeves. Long pants and sneakers are preferable to shorts and sandals. A protective apron will help prevent ruining clothes from oil spills, as well as be a first defense in the event of a lye spill.

Equipment and Supplies Necessary for the Safety of the Product

Most of the equipment necessary for making liquid soap can be found in the kitchen. Stainless steel pots, slow cookers, plastic pitchers, glass measuring cups, mason jars, cooking thermometers, whisks, spatulas and stick blenders are the common tools of the trade. Soapmakers are very adept at repurposing and making do with what they have on hand. The soapmaker must have the following materials in order to make liquid soap safely: 1) a good set of scales 2) phenolphthalein 3) MSDS of all raw materials and 4) a batch code sheet for every batch of soap you make.

Scales: Just because soapmakers have been making soap for centuries without the benefit of a set of accurate scales is no reason to opt out of this necessary piece of equipment. If the age old term "lye soap" is any indication of the often resultant product, I rest my case.

There are plenty of unknowns and variables in the art of handcrafted liquid soapmaking and there is no reason to add any additional burdens upon the soapmaker. Accurate scales along with good

note taking skills allow the soapmaker to replicate successes and avoid failures. It also goes a long way in avoiding "lye" soap!

Good scales can be purchased rather inexpensively. In purchasing scales, the soapmaker should look for a set that:

- Is appropriate to the amount of material being weighed. They should be able to weigh the heaviest item you plan to weigh within its maximum weight capacity limit. They should also be able to accurately weigh the smallest item you plan to weigh.

    If you need to weigh 15 grams of a preservative and your scale has a minimum accuracy of 1 ounce, you need a scale with a smaller readability, such as 1 gram. If you need to weigh 72 ounces of olive oil and your scale's maximum weight capacity is 2 pounds, you need a scale with a larger weight capacity. You may need more than one scale to accommodate all your weighing needs.

- Has a tare feature. Tare allows the soapmaker to reset the display to zero when they place a container on the scale. It also allows the soapmaker to reset the display to zero between additions of each ingredient, simplifying the process exponentially.

- Has the ability to disable the automatic shut-off feature. It is very frustrating to have the scale shut off automatically while opening a new container of olive oil in the middle of weighing a large batch of soap.

- Has a removable platform for ease of cleaning. Soapmaking is messy! Trying to clean the platform while still attached to the scale can damage the scale if too much pressure is applied.

- Has an A/C adaptor. Batteries should only be used for back-up. Low batteries can result in inaccurate measure, not to mention failing in the middle of weighing a batch.

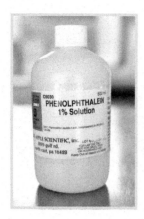

Phenolphthalein: Ensuring the soap is free from excess alkali is an essential part of good manufacturing practices. Creating a bright, clear soap free from excess fatty acids is an indication of a skilled soapmaker. Both can be accomplished with the use of phenolphthalein, a chemical compound used as a pH indicator. Phenolphthalein remains clear when added to acidic solutions and turns pink in overly alkaline solutions. It is excellent in indicating the neutral point between a weak acid and a strong base, which is exactly the reaction the soapmaker creates through saponification. In titrating alkali soap, the color slowly changes over a pH range of 10.0 to 8.3, from pink to clear, indicating the point at which the soap is neutral. Phenolphthalein is inexpensive and can be purchased from most chemical supply houses pure in its crystalline form or as a 1% solution ready to use.

MSDS Sheet: MSDS (short for Material Safety Data Sheet) should be provided to the soapmaker by the supplier. It outlines any potential dangers of the product with guidelines on how to handle the product in a safe manner as well as what to do in the case of any accident. MSDS sheets should be accessible to all workers and emergency personnel at all times.

Batch Code Sheet: A batch code sheet documents the production of the soap from the suppliers of the raw ingredients to the finished product. It should include the manufacturer or supplier of the raw ingredients with the lot or serial number of the ingredient, the list of ingredients, the process of manufacture and any variations or observations in the soapmaking process as well as the name of the soapmaker and the date of manufacture. The batch code sheet serves to assure the product was made according to good manufacturing procedures. It also serves the soapmaker as a reference in the event something goes wrong with the batch. Was it a new supplier, a new batch of olive oil, a mismeasure of ingredients? Those questions can easily be solved by keeping a good batch code sheet. See Appendix A for a sample batch code sheet.

Safety requires vigilance and a conscious decision to make safety measures a habit. It also requires having an understanding of the raw materials and soapmaking procedures.

CHAPTER

INTRODUCTION TO POTASSIUM SOAPS

Soap Pastes, Liquid Soaps, Jelly Soaps and Soft Soaps

"... the distinction between hard (or soda) and soft (or potash) soaps. The soda salt, on exposure to air, will gradually lose nearly all its water, at a rate depending upon the dryness and temperature of the air. The potash salt, on the other hand, never becomes thus dry, but retains enough water to form a soft slimy jelly..."

~ *Soap Candles Lubricants and Glycerin*
(CARPENTER & MUSK, 2ND EDITION, 1895)

Liquid soap, soft soap, soap paste, oil soap, jelly soap and cream soap are all potassium based soaps. Made by mixing an aqueous solution of potassium hydroxide with fats and/or oils, potassium soaps are considered to be the simplest soaps to make. Simple does not always equate to easy. Before jumping right into the potassium soap pot, it's important to obtain a basic introduction of the soapmaking process, the nature of some of the raw materials and perhaps dispel some of the notions held by other soapmaking methods.

The Chemistry of Soap

Soap is a chemical reaction. While the typical handcrafted soapmaker doesn't need to delve too deeply into the chemistry of soap, there are some aspects they need to understand. Specifically, the soapmaker needs to know what is going on in the soap pot as well as what not to do with what is going on in the soap pot in order to make good soap safely. A few chemistry terms will help the handcrafted soapmaker speak the language practiced and chemistry methods will help the soapmaker do what is already done: Measuring, stirring and experimenting a little more precisely.

Lipids, Triglycerides, Fatty Acids and Glycerol

Lipids are another name for the fats, waxes and oils we use to make soap. Fats and oils are made up of triglycerides (three molecules of fatty acids attached to one molecule of glycerin or glycerol), small quantities of free fatty acids (fatty acids not attached to glycerin) and unsaponifiable plant matter (natural substances within the plant that cannot be made into soap). Waxes contain fatty acids but no glycerin.

Saponification

When fats or oils are mixed with an alkali solution, the triglyceride is split into glycerol and fatty acids. Saponifcation is the chemical reaction of the fatty acids with a strong base or alkali (such as potassium hydroxide) to create a salt of the fatty acid, what we commonly refer to as soap. Free glycerin is produced as a byproduct of the saponification reaction.

pH

The pH of soap is perhaps the industry's most confusing, controversial and misinformed topic. Scroll the internet and you will find claims of handcrafted soap with a pH of 7, directions on how to rebatch handcrafted soap to achieve a pH of 7 or questions on how to lower the pH of handcrafted liquid soap to 5.5 for a face wash. It's a mess!

Soap, whether handcrafted or commercial, has a pH range around 9–10.1. Anything over is what you would call lye soap and anything under is a detergent or the result of a bad test strip. In resolving the pH riddle, understanding pH, knowing what it measures, how it is measured and why the chemical reaction of the soapmaking ingredients cannot be manipulated to the pH of the competing surfactants should be answered.

pH is the measure of the acidity or alkalinity (basicity) of a solution on a scale of 0–14 with 7 being neutral. Any solution with a pH less than 7 is considered acidic. Any solution with a pH greater than 7 is considered alkaline or basic (alkaline and base are terms used interchangeably). The lower down the pH scale the stronger the acid and the higher up the pH scale the stronger the alkali. Distilled water with a pH of 7 is neutral. Comparing the pH of acids and bases in solution of equal concentration indicates the relative strength of the acid or base. Hydrochloric acid is a strong acid with a pH of 1.1. Acetic acid (vinegar) with a pH of 2.9 is considered a weak acid. Sodium and potassium hydroxide are strong bases with a pH of 13. Sodium bicarbonate with a pH of 8.4 and sodium carbonate with a pH of 11.6 are considered weak bases. Here comes the part that will help the soapmaker understand why the pH of soap is alkaline.

- When a strong acid is reacted to a strong base of equal strength, the acid and base neutralize each other and the pH will be 7.

- When a weak acid is reacted to a weak base, the solution will be neutral if the acid and base are of the same strength. If the acid is stronger the pH will be acidic. If the base is stronger the pH will be basic.

- When a strong acid reacts to a weak base the pH will be acidic. How acidic depends on the relative strength of the acid and the base.

- When a weak acid reacts to a strong base the pH will be basic. How basic will depend on the relative strength of the acid and the base.

> Acids and bases should never be combined if the resulting reaction is unknown. The consequence of doing so could cause an explosion or the release of toxic gases.

The recipe nature gave us to create soap is the combination of a weak acid (fatty acids from the oils of plants and animals) mixed with a strong base (potash from the ashes of hardwood trees or ashes of seaweed to make soda ash). When the fatty acids from the oils are mixed with the alkali in solution they come together in a chemical reaction known as saponification. Since it is a reaction of a weak acid with a strong base, the resultant soap will have a pH higher than 7. The exact pH depends on the pH of the fatty acids in the mix. Coconut oil soap will have a higher pH than olive oil soap, but all soaps will be within a pH of 9.0–10.1. Attempting to artificially lower the pH will change the chemical structure and it will no longer be just soap.

Neutral Soap

This is where some of the confusion lies. Repeatedly throughout the book references will be made to assure the soap is neutral. After going on ad nauseam regarding the impossibility of achieving a neutral pH in soap, why turn around and talk about soap being neutral? Everyone thinks of neutral as the measure of pH, as in the pH of water, just as everyone thinks of fluid oz. as being the weight of water. If the soapmaker changes mind set and considers neutral as being subjective to the object measured, soap may be considered as neutral if it no longer contains excess oils or lye, despite the fact it is still alkaline. (E.g. a gallon container of glycerin weighs approximately 10 lbs. versus a gallon container of water which weighs approximately 8 lbs., but both are considered a gallon by fluid measure despite the difference in weight). Therefore, a soap that has no excess fatty acids or alkali left in solution can be defined as being a neutral soap. Achieving a neutral soap is one of the most difficult aspects of liquid soapmaking.

Phenolphthalein

The best way to determine whether any excess alkali still remains in the soap is with an acid/base indicator. An acid/base indicator is generally a solution of a weak acid or a weak base that indicates the degree of acidity or basicity of the object being measured by a change in color. It is also used to indicate the completion of a chemical reaction, the point at which the alkali and acid are in balance. There are numerous acid/base indicators and the choice of which one is right for the job at hand is dependent upon the range of pH the acid/base indicator measures. Phenolphthalein is the indicator of choice for the handcrafted soapmaker. Its color changes over an approximate pH range of 8.3–10.0, the point in which half of the indicator is in its acid form and the other half in the form of its conjugate base. Anything below this range the solution will be colorless. Anything above this range the solution will be pink to bright fuchsia. Testing for alkalinity is as easy as placing a couple of drops of phenolphthalein in a solution of liquid soap. If the color remains the same, the soap is either neutral or contains excess fatty acids. If the solution turns pink, there is excess alkali in the soap and it will require neutralization which will be discussed more completely on page 30.

Phenolphthalein is the best method of testing for excess alkali in soap. Unlike pH test meters, which are expensive, must be calibrated between tests and only test for a specific pH, phenolphthalein is easy, inexpensive and indicates the point at which the chemical reaction is complete. Soap's pH is dependent upon the oils used in the formula so it has a pH range, not a specific pH.

Potassium Hydroxide

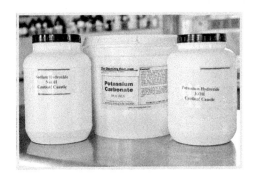

Potassium hydroxide is the main alkali used in the production of liquid soap. It is one of the most dangerous chemicals on the handcrafted soapmakers ingredient list as is sodium hydroxide. As a strong base, it has the potential to cause serious harm and even death. Regardless of how accustomed the seasoned soapmaker becomes in working with it, potassium hydroxide must always be treated with the greatest respect. Knowing how to handle potassium hydroxide and what to do in the event of an accident will greatly reduce the risks associated with its use. Examination of the MSDS sheet will explain all the necessary precautions the soapmaker should take as well as what to do in the event of an accident.

Potassium hydroxide, also referred to as caustic potash and KOH (its chemical formula), indicates the solid form which comes in flakes or beads. Solutions of potassium hydroxide are usually referred to as caustic lye, potash lye or potassium hydrate indicating their liquid state. The liquid form is most usually purchased as a solution of 45% potassium hydroxide in 55% water. Because transportation, storage and handling of the liquid form are problematic, most handcrafted soapmakers choose to use the solid form. Potassium hydroxide flakes are especially recommended over the beads if the soapmaker uses both sodium hydroxide and potassium hydroxide in soapmaking. The difference in the appearance of the two will allow the soapmaker to tell at a glance if the correct hydroxide is being used.

Because the molecular weight of potassium hydroxide is heavier than sodium hydroxide 56.1 grams/mole vs. 40 grams/mole, (Mole is a standardized unit of measure used in chemistry for expressing amounts of a chemical substance) it requires more potassium hydroxide to saponify the same amount of fats and oils as sodium hydroxide. Confusing the two has the potential of creating a dangerously lye heavy cp soap or a heavily super-fatted liquid soap, wreaking havoc with the batch either way. All references and recipes in this book are directed toward the solid form.

Potassium hydroxide is used in making liquid soap because it is more hygroscopic; meaning it easily absorbs more water than sodium hydroxide. For that reason, potassium soaps are softer and more soluble than sodium or bar soaps. Its high affinity for water means it takes less water to liquefy; therefore the finished product carries more cleaning power in liquid form, is more likely to remain in solution and will rinse off easier.

Its hygroscopic nature also means it is very difficult, if not impossible to purchase pure potassium hydroxide. Most suppliers guarantee it to be anywhere from 95%–99.9% pure. The purity however is not a guarantee of its strength, but indicates the absence of other chemicals or minerals in the mix. Unless the potassium hydroxide is produced and filled in vacuum, such as the case of lab grade potassium hydroxide, it will absorb the moisture as well as the carbon monoxide in the air as it is being processed. The moisture dilutes its strength by absorption (think of a sponge soaking up water) and the carbon monoxide reacts with the potassium hydroxide to form potassium carbonate, a weaker alkali. To compound matters, once purchased, every time the container is opened, it absorbs the moisture as well as the carbon monoxide in the air. The handcrafted soapmaker is left to speculate about what exactly is in the container. This is one of the reasons the amount of potassium hydroxide in a formula will appear high. We will discuss other reasons further on.

When mixed with water potassium hydroxide creates an exothermic reaction, meaning it generates heat. It generates more heat and energy than its counterpart sodium hydroxide as evidenced by the rather unsettling "whooshing" noise it makes upon stirring. This is normal and is no cause for undue concern; however it should serve as a constant reminder of the serious need to treat it with respect. Nonetheless, the reaction is rather short lived and when mixed with oils, the caustic lye stubbornly languishes in the pot unless coaxed into saponification. For this reason the hot process method is the soapmakers method of choice because more heat is required to saponify the oils with potassium hydroxide.

Sodium Hydroxide

Sodium hydroxide, with the molecular formula NaOH and also referred to as caustic soda, is most often used in creating solid bar soaps. Attempting to make a liquid soap entirely with sodium hydroxide as the alkali will result in a snotty, stringy, gelatinous mess unless diluted to the point of being ineffective. However, a percentage of sodium hydroxide in conjunction with potassium hydroxide will help increase viscosity and boost lather. It will also increase the energy of the alkalis, especially helpful when working with formulas containing additives or higher amounts of water. More energy equates to more complete saponification and often much shorter cook times. Up to 33% sodium hydroxide can be substituted for the potassium hydroxide. Keep in mind sodium hydroxide is lighter and therefore requires less by weight than potassium hydroxide to saponify the same amount of a given fat or oil. Calculate the amount of sodium hydroxide following the directions in "Chapter 15: How to Formulate Liquid Soap Recipes," or use a reliable lye calculator which can be found online. Recommendations for reliable lye calculators can also be found in "Chapter 15."

Potassium Carbonate:

Potassium carbonate, with the molecular formula K_2CO_3 is also referred to as potash or pearl ash. It is a milder alkali than potassium hydroxide with a pH of 11.6. While it does not have the strength to react completely with the oils to form soap and glycerin, it will react with the free fatty acids or trace metals found in the soapmaking ingredients to bind them into soluble and insoluble soaps thereby contributing to the transparency, brightness, detergency, consistency and shelf life of the soap. Another advantage of potassium carbonate is its ability to keep a concentrated soap fluid. By inserting itself between the soap molecules, potassium carbonate will loosen a stiff paste to make it easy to stir. With additional water, potassium carbonate will allow the process to go directly into a liquid or gel state, eliminating the need for a paste entirely. Because of its alkalinity, additional neutralization may sometimes be required when using potassium carbonate in a formula. Always add potassium carbonate to liquids at room temperature or lower. Potassium carbonate foams easily and generates an exothermic reaction, milder than potassium hydroxide, but one that can volcano if added to hot liquids.

Saponification Value

Theoretically it should be possible to take the saponification value (the amount of potassium hydroxide required to saponify a particular fat or oil); and with a little arithmetic, arrive at the amount of potassium hydroxide required to perfectly neutralize the given oil. The first problem is of course the unknown strength of the potassium hydroxide, the second is the fact the saponification value is a range of values, not a constant.

    A trip to the grocery store will convince you all olive oil, or any other oil for that matter, is not created the same. Geographical location, soil conditions, oil extraction methods and fluctuations in the weather affect the chemical composition of the oil-giving plant (or

> Always ask for a specification sheet when ordering soap-making oils and fats. The specification sheet will include the saponification value along with other relevant information specific to the oil or fat being ordered.

animal) from one harvest to the next which in turn affects the saponification value.

Using olive oil as an example, the range of the saponification value on any given harvest can require from 184 to 196 milligrams of potassium hydroxide to completely neutralize one gram of olive oil. The value of any given olive oil should be in that range, but the specific value for the oil in the soap pot is relatively unknown.

Most suppliers, lye calculators, soapmaking websites and books take an average of the minimum and maximum values, convert the value to a percentage and refer to it as the saponification value. This is the value with which most soapmakers are familiar. Aside from being misleading, it is also confusing and has led many in the handcrafted soapmaking industry to believe the value being expressed is precise. This misconception has the potential of giving the handcrafted soapmaker a sense of overconfidence in his/her ability to produce neutral soap.

The difficulty of achieving a neutral liquid soap by calculating an unknown strength of lye with an average range of values is apparent. Large manufacturers test every ingredient that comes through their doors to determine the specific value and quality of each ingredient. It makes sense for them to do so in light of the vast quantities of materials involved in the operation. The handcrafted soapmaker, making extremely small batches, buying in small lots without the luxury of additional man-power must find a way around the unknown in order to be efficient, yet accurate. The answer out of this dilemma is to prepare an over-alkaline soap and neutralize the alkalinity at the end of the process by various methods. How much is too much and how much is not enough?

Let's take for example some of the variables that may apply.

- Strength of the potassium hydroxide in relation to its grade: Food grade potassium hydroxide will have greater strength than technical grade.

- Climate: The soapmaker living in arid Arizona will have fewer problems than someone living in humid Florida.

- Storage and usage: The last half of a fifty pound bag has the potential of being considerably weaker than the first half depending on how often and how long the bag is left open.

- The specific soap being produced.

With all these variables it would be difficult to determine a single value that would conveniently address all situations. We can therefore create a table in which the soapmaker can determine which value would best serve his or her purpose.

Experimentation has led me to conclude the following without warranty while using technical grade potassium hydroxide:

1. Using an excess of 8% alkali with fresh (meaning recently purchased and just opened) potassium hydroxide will achieve a clear soap with little or no excess alkali most of the time.
2. Using an excess of 10% alkali with the same criteria as above will achieve a slightly alkaline soap that will require neutralization most of the time.
3. Using an excess of 10% alkali with potassium hydroxide that has been exposed to the atmosphere numerous times will achieve a clear soap with little or no excess alkali most of the time.
4. Using an excess of 12% percent alkali with potassium hydroxide that has been exposed to the atmosphere numerous times will give me a slightly alkaline soap that will require neutralization most of the time.

Oil	SAP Value	KOH Average Value @ 0%	NaOH Average Value @ 0%	KOH Alkali Value @ 8%	KOH Alkali Value @ 10%	KOH Alkali Value @ 12%
Castor Oil	176–187	181.5	129.4	196	199	203
Coconut Oil	248–265	256.5	182.8	277	282	287
Palm Oil	190–209	199.5	142.2	215	219	223

As you can see in the table, the alkali values are considerably higher than the average values. To someone making cold process soap the numbers would appear scary to the point of chemical burns without understanding the reason behind the inflated values.

If the liquid soapmaker desires to calculate the precise saponification values for the specific hydroxides and oils they are using in their soap pot, *Scientific Soapmaking* (Dunn, 2010) provides complete directions for the process. Otherwise making an alkaline liquid soap and then neutralizing the excess alkali out of the soap is the best method for the handcrafted soapmaker working in small batches.

Water

Water is the solvent that carries the potassium hydroxide to the fats and oils. It also determines the concentration and clarity of the resultant soap. In essence, potassium soaps are water based, meaning the majority of ingredients by weight is water. Soft or distilled water is essential in producing soaps which clean well and remain clear. Minerals in hard water combine with the lye solution and make insolvent mineral soaps (also known as soap scum) which lessen the cleansing effect and restrict clarity.

Oils and Fats

The goal of a finished liquid soap should be a neutral, transparent solution with a rapid lather, good detergent action, agreeable odor and satisfactory viscosity. Transparency, lather, detergent action, odor and to a large degree the viscosity of the finished product hinges mainly on the soapmaker's choice of oils and fats.

No single fat or oil provides the perfect criterion, so we must look at the makeup of the oils and fats in order to find a combination that will give us our very own magical formula.

The selection of fats and oils are dependent primarily on the fatty acid makeup of the fat or oil. It is the action of the potassium hydroxide upon the triglycerides that form soap and glycerin. All oils contain various mixtures of fatty acids. The most common fatty acids that concern the soapmaker are lauric, myristic, palmitic, stearic, oleic, linoleic, linolenic and ricinoleic. Refer to the table in Appendix D for a list of common soapmaking oils and percentages of their fatty acid makeup.

Fatty acids can be divided into saturated (lauric, myristic, palmitic and stearic) and unsaturated fatty acids (oleic, linoleic, linolenic and ricinoleic). Saturated fatty acids are generally referred to as fats and unsaturated fatty acids as oils.

Saturated fatty acids are straight chain fatty acids, which mean they have all the hydrogen that the carbon can hold and therefore have no double bonds. Because they are so dense, they are usually solid at room temperature, resist rancidity and have a higher melt point than unsaturated fatty acids. With the exception of lauric and myristic acids, saturated fatty acids will cloud liquid soap if used in large measure. As a small

percentage of the formula, they can help to stabilize the lather and increase the viscosity of the finished soap. Lauric and myristic acids, because of their shorter chain length and lower molecular weight, are atypical of saturated fatty acids in the fact they produce abundant fluffy lather, are very soluble, produce great transparency, have excellent cleansing and antimicrobial action. They can also be mild irritants.

Old time soapmakers used saturated fats in their soft soap during the summer to prevent weeping in the soap due to rising temperatures. In the winter, they were used to "fig" or "rice" the soap. Figging and ricing were a result of the stearic acid in the saturated oils separating out into snowflake or rice like patterns of white soap in the otherwise transparent soft soap and was considered a sign of superior soapmaking skills. Lard was used to make a product called silver soap, similar today to pearl soap and the front runner to cream soap.

Unsaturated fatty acids have one or more double bonds connecting their carbons; further classification divides these unsaturated fatty acids into monounsaturated (one double bond) and polyunsaturated (more than one double bond). Oleic and ricinoleic acids are monounsaturated fatty acids and linoleic and linolenic acid are polyunsaturated fatty acids.

Unsaturated oils commonly referred to as soft oils provide emollience and viscosity to liquid soap. They give a close, rather slow and somewhat weak lather. They should be chosen for their stability and price.

- **Stability:** Soft oils high in oleic acid such as olive, high oleic sunflower, high oleic safflower, apricot kernel, sweet almond, avocado and macadamia nut are more stable than oils with high linoleic and/or linolenic fatty acids such as corn, soybean, hemp, grape seed and evening primrose oils.

- **Price:** Price should be a major consideration when choosing soft oils for liquid soapmaking. They generally make up a large percentage of the oils in the formula and are most often the most expensive. Oils with similar fatty acid profiles will give comparable results with little variation between viscosity, lather, emollience and transparency regardless of the price. A few things to consider when choosing soft oils:

  - If color matters, the darker the oil, the darker the soap. Olive oil will make a deeper amber soap than sweet almond or safflower which are lighter in color.

  - Rice bran oil due to its high wax content has a tendency to cloud in liquid soap. Avocado oil with its high degree of unsaponifiables can also cloud. They should be used as a smaller percentage of the formula.

  - Oils that contain a higher percentage of linoleic acids, even though they also contain high percentages of oleic acid have a tendency to make a stronger smelling soap. Oils such as canola, peanut, sweet almond and apricot kernel should be used as minority oils in the formula.

Ricinoleic acid, derived from castor oil, is different than its other unsaturated fatty acid counterparts in that it is a monounsaturated fatty acid with a hydroxyl (OH) group at a specific position. This hydroxyl group protects the double bond by preventing the formation of peroxides, making it more stable and giving it solubilizing and plasticizing properties. Due to its solvent nature, it has great cleansing properties and is a humectant and emollient.

Butters and waxes are used for the same purpose as saturated fatty acids and also for the wealth of nutrients they provide to the soap due to the larger percentage of unsaponifiables. Because waxes repel water more efficiently (hydrophobic) than oils and fats, they require a much longer cook time in liquid soap. Cocoa butter is the easiest butter to incorporate into liquid soap. On the other end of the spectrum is shea butter. It is especially notorious in this regard and experience has shown it works better with the paste method of liquid soap rather than any of the other methods outlined in the book.

Blending Oils: One of the soapmaker's greatest skills is in manipulating the oils in order to get the desired result. It is a balancing act to be sure, but a few simple tips on how certain oils perform in the liquid soap pot should help shorten the learning curve. The two most featured oils are coconut and castor. They are indispensable in the soap pot in combination with other oils.

Coconut oil with its cleansing action and ability to produce a quick abundant lather is the foundation of most liquid soaps. Coconut oil's enthusiastic desire to please the liquid soapmaker is endearing, but often borders on the overzealous side of soapmaking.

- It is one of the quickest oils to saponify, but goes ballistic at higher temperature, often crawling up, out and over the sides of the pot alarmingly fast!

- It continues past trace effortlessly with little worry of separation and forms a solid lump of paste that is more conducive to slicing than stirring.

- It is highly soluble and will remain fluid up to a 41% soap concentration without reverting to paste. It is so soluble it dissolves to a consistency comparable to water even in its highest concentration.

- It produces the quickest, most abundant lather of shortest duration.

- It garners the highest honors for its cleaning ability but can leave the skin feeling dry or irritated.

For the reasons cited above it is a fundamental part of almost any liquid soap formula, yet rarely is it used alone. Other oils are blended to add moisture, body and stable lather to the soap.

Castor oil is a soft oil that stands apart from other oils in soapmaking and like coconut oil, deserves special attention. Contrary to the widely held belief castor oil increases lather if used as a small percentage of the formula, I have found it not to be the case. In trialing numerous recipes for the purpose of this book, I have consistently found the addition of castor oil negatively impacts the lathering ability of the soap. The increase in lather when castor oil is incorporated into a cold process bar soap is most likely due to the castor oil making the soap more soluble and easier to generate more lather with less effort but more soap. Lather isn't everything and castor oil plays a very important role in liquid soapmaking due to its unique structure, more so than in any other type of soapmaking. The benefits of using castor oil in liquid soap include the following attributes.

- It has unusual emulsifying properties.
- It will improve the clarity of liquid soap.
- It is the quickest of all oils to saponify and will aid in correcting a soap that is too alkaline.
- Coconut oil soap with 20% castor oil will make a soap that will remain clear at lower temperatures.
- Using a portion of castor oil in liquid soap will permit up to a 48% soap concentration.
- A soap formulated of 100% soft oil will dissolve easier with the addition of castor oil.

Skin care: In our quest for creating the perfect soap we are so intent in our examination of the oils we use to assure clarity, lather, emollience and stability, we often overlook other attributes they bring to the equation. Specifically, we overlook, the natural substances within the oil giving plant that cannot be made into soap and are referred to as unsaponifiables. Unsaponifiables are the hydrocarbons, fatty alcohols and pigments that make up in large part the nutrients of the oil or fat which make our customers keep coming back for the wholesome goodness our soaps impart. A list of some of the most common unsaponifiables, with their nutritional benefits in the oils they are most prevalent in highest concentration is below.

- **Triterpenoids:** This group includes squalene, steroids and sterols.

    - **Squalene** is a metabolic precursor to steroids and sterols and is found in the sebum of human skin. It is used in cosmetics as a natural oil-free moisturizer. Squalene is found in olive, wheat germ, rice bran and fish.

    - **Sterols & Steroids:** A common steroid, cholesterol, is found in the human body. Its counterpoint in the plant kingdom is phytosterols which are known to help reduce cholesterol in humans. Sterols are critical components of cellular membranes and they serve as a forerunner to many hormones. In skin care they reduce aging by their ability to hold water in the skin cells. Sterols are found in higher concentration in sesame, canola, corn and evening primrose with lower concentrations in peanut, safflower, soybean, borage, cottonseed, coconut, palm, olive and avocado oil. Lanosterol is a natural constituent of wool fat.

- **Vitamin E:** This group includes tocopherols and tocotrienols

    * **Tocopherols,** mainly alpha-tocopherol, are important antioxidants that help protect the cells against free radicals. Used topically they help the skin look younger, promote healing, reduce scar tissue and reportedly aid in the treatment of eczema, cold sores, skin ulcers and shingles. Tocopherols are found in highest concentrations in canola, cottonseed, olive, peanut, safflower, soybean and sunflower oils

    * **Tocotrienols** appear to be the most valuable form of vitamin E. Tocotrienols penetrate rapidly through the skin and help prevent aging and skin damage by oxidative rays. Because they are more potent than tocopherols, they neutralize free radicals at a faster rate. The richest source of tocotrienol is palm, palm kernel, coconut and rice bran oils.

- **Carotenoids:** Carotenoids are a primary source for vitamin A in humans and are said to have antioxidant properties similar to vitamin E. Retinol, a derivative of beta carotene, is noted for its ability to treat skin disorders including psoriasis and acne. Oils rich in carotenoids include virgin palm oil, pine nut oil and carrot oil.

Glycerin is a product of the saponification process and remains in the soap unless removed for other uses. Adding additional glycerin to the soap helps moisturize the skin by attracting moisture in the air. It increases the wetting ability of the soap, provides clarity, speeds saponification and allows for a more concentrated soap solution by its solvent action. Some of the procedures outlined in this book require a solvent to carry the process forward. While there are many solvents available, glycerin is the most natural and obvious choice.

CHAPTER

HOW TO MAKE LIQUID SOAP USING THE PASTE METHOD

"Soft soaps are more easily made than hard soaps. The fats and oils are boiled with the alkali till the saponification is complete; then the soap is made, and only needs to be run into firkins or tin canisters to be ready for sale."

~ *Textile Soaps and Oils* (HURST AND SIMMONS, 1914)

Step by Step Instructions

It may appear placing the directions before the recipes is synonymous with placing the cart before the horse, but it is important to understand and review the steps prior to gathering the materials and arranging the workspace. Also, the steps to making soap pastes, liquid soaps jelly soaps and soft soaps are fairly straight forward with minimal variations. Providing in depth directions with each recipe would be extremely repetitive. Placing the general directions at the beginning of the chapter should allow for easier reference for all recipes.

Step 1 — Safety First!

Step one is always the safety step! If you haven't yet read the beginning chapter of this book, "Safety First" this would be a good time. As a reminder: Don your safety goggles, gloves and protective mask and clothing; arrange your workspace efficiently; make sure you have sufficient ventilation and remove or eliminate any distractions.

Step 2 — Gather Thoughts and Materials

Before beginning any project, make sure all the necessary materials and equipment to complete the project are available. Carefully read over the recipe and the following directions before beginning. Lay out

the equipment and gather the supplies together making sure there is adequate work space. Now you are ready to begin!

Step 3 Prepare the Lye Solution

The first step in creating a lye solution is the water portion. The water should be soft or distilled and should be at room temperature or lower. The vessel should be of heavy plastic or stainless steel. Polypropylene beverage pitchers with snug fitting lids and pourable spouts found at discount stores are a good choice. Make sure the container is large enough to accommodate the water and the alkali. Weigh the water and set to the side.

All of the recipes found in this book contain potassium hydroxide. Some recipes will also contain sodium hydroxide. Additives unaffected by the lye solution may also be added at this time, but instructions in the specific recipe will direct you.

> All alkalis are corrosive to most metals, even in dilute concentration. Reaction with aluminum, tin or zinc (galvanized) or any alloy containing these metals create flammable hydrogen gas, therefore all equipment used in soapmaking should be of non-reactive materials.

Weigh the required alkalis into containers made of heavy plastic or stainless steel. The alkalis can be measured and added one at a time into the water or measured in the same container (taring the scale between each measure) and added to the water.

Slowly add the alkalis to the water with gentle stirring to avoid splashing. Continue stirring until all are dissolved and the solution is clear. Cover the lye solution with a tight fitting lid and set aside. Preparing the lye solution first gives it time to cool down. Optimal temperature of the lye solution before mixing with the oils should be between 135°–165°F (57°–74°C).

> Using glass containers to mix lye solutions will cause the glass to degrade over time which may bring about sudden and unexpected breakage.

> Always pour the alkali into the water! Never pour the water into the alkali! Doing so may cause a volcano effect which may result in severe injury to you and/or damage to the surrounding area.

Step 4: Weigh and Melt the Oils

If a large portion of the recipe calls for coconut oil or other saturated fats or waxes, weigh them first and place them in the cook pot before weighing and adding the soft oils to the mix. The fats can begin melting while measuring out the soft oils. Once the fats are melted, add the premeasured soft oils to the soap pot. This will decrease or eliminate any cool-down time.

Step 5: Bring the Combined Oils to the Desired Temperature

Depending on the mixture of oils, it is necessary to either heat the oil mixture or allow it to cool to the recommended temperature. Recipes higher in coconut oil require lower temperatures (160°–175°F [71°–80°C]) than recipes higher in soft oils (170°–185°F [77°–85°C]).

Step 6: Stir the Lye Solution into the Oils

With a wire whisk or spatula, slowly stir the lye solution into the fats and oils. Check the temperature of the solution. If it is below 160°F (71°C) return the soap pot to the heat source and allow the solution to reach the desired temperature. If the temperature is above 185° (85°C) allow the solution to cool down before continuing. Stir until the oils begin to form an emulsion with the lye. Now (unless you really want to spend some quality time with your soap pot) is the time to exchange your whisk or spatula for some type of mechanical mixer. Liquid soap is notoriously slow to achieve trace. For small batches (under twenty five pounds [11.33 Kg] of oils) the stick blender is the mixer of choice. It is easy to handle and aids in the formation of a quick emulsion.

Maintain the optimal temperature while alternating stick blend-

Tip Glycerin may be added to the soap at any time to give emollience and boost the lather of the soap. Adding it as a part of the lye water or discounting the lye water and adding it at this time will greatly speed up the process.

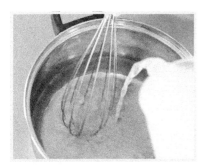

ing with manual stirring, being careful not to allow the soap stock to rise above the maximum temperature. Potassium soaps may languish for hours without sufficient heat, but excessive heat may cause the soap to swell, puff up and boil over as the chemical reaction between the lye and the oils generate additional heat to the pot.

Step 7 — Continue to Trace

Trace is generally defined as the point at which the emulsion thickens sufficiently that a portion of the solution when dribbled over the surface leaves a trail.

Step 8 — Liquid Additives

Once the soap emulsion has begun to trace is the time to add any additional liquid additives. Adding additives at this time will allow them to be fully incorporated before the emulsion becomes too stiff. They should be at or near the temperature of the soap emulsion when being incorporated.

This step is only necessary with recipes containing other liquids such as aloe vera. If not adding other liquids, skip this step and proceed to the next step.

Step 9 — Continue Past Full Trace

The definition of full trace in potassium soap differs from that of sodium soaps. Generally, the soapmaker will notice the emulsion thickening into the typical creamy or pudding like consistency familiar to cold process soap. Continue past this stage until what is referred to as the "taffy stage" is achieved, in which the soap takes on a taffy-like appearance. It is during this stage the soapmaker needs to watch for separation. Separation occurs often enough, especially with a majority of soft oils and presents as an applesauce appearance to the soap formula. Depending on the oils, full trace is achieved only when the stock becomes extremely thick. Soaps with a large majority of soft oils will resemble jelly and stirring is possible throughout the process. Soaps with a majority of coconut oil, will resemble school paste and will be so stiff it can only be jabbed at and mashed! Quitting before reaching this stage may cause separation during the cook.

Separation

Separation can be severe or minor. If severe, the soap paste will have a curdled or applesauce appearance and will need to be vigorously stirred with whatever apparatus will allow a healthy stir (stick blender, whisk or potato masher depending on the stiffness of the paste) until the paste goes back into a homogenous blend. The soap will need to be checked every twenty minutes and stirred back into emulsion until it no longer separates. If the separation is minor, the lye solution will separate down to the bottom as a clear liquid. If no liquid is present, but the paste easily slides around the bottom of the pot as if it had been well oiled, the separation is very slight. Slight separations are more easily incorporated. Patience in the beginning will save time and worry later on in the process.

> The answer to separation is always more stirring until there is no longer a separation!

Step 10 › Cook the Soap

Once the soap has reached a paste or jelly like consistency, it is necessary to continue cooking the soap mixture in order to fully saponify the fats and oils. As the cooking continues, the paste or jelly will develop a translucency, often referred to as the Vaseline stage. This is a sign that all the fatty acids in the oil have combined with the lye and the cooking process is complete.

A variety of cooking methods are available and soapmakers have been very ingenious in this process. Methods employed include double boilers, crock pots, turkey roasters, ovens and other similar methods. Whatever method the soapmaker chooses, the key is in creating steady even heat to the paste at temperatures between 180°–195°F (82°–91°C) for two or more hours until the soap reaches translucency. The cook time will vary depending on several factors including the type of oils, degree of heat as well as the quantity of soap.

Whatever method employed, it is important to avoid excess heat and evaporation. Excess heat may cause the soap to swell and boil over the pot. It also oxidizes the soap creating an off odor and color. Excess evaporation causes the soap to be firmer and more difficult to stir, thereby making it more prone to scorching and oxidizing.

Directions to the following processes will help guide the soapmaker in selecting the method of choice or in creating one's own.

Double Boiler Process: A double boiler is a set of two nesting pots or pans designed to allow slow even cooking of the upper pot by the action of the boiling water in the lower pot. A double boiler may be constructed using any pot that will easily fit inside the other with room around the edges for the water to lap up the sides of the upper pot. The upper pot should rest in the water, not touching the bottom of the lower pot. The handles of some soap pots are wide enough to prevent it from resting on the bottom. When that doesn't work, placing the soap pot on canning jar lids or a tripod made specifically for this purpose keeps the soap pot elevated.

To proceed, fill a pot large enough to accommodate the soap pot with three to four inches of water and bring to a boil. This should be done prior to preparing the soap paste. When the soap pot full of soap paste is placed in the lower pot, the water should lap up the sides to the level of the paste in the soap pot. This will ensure the paste cooks evenly. Adjust the water level accordingly.

Cover both pots with a lid or aluminum foil to reduce steam to the room and prevent unnecessary evaporation of the boiling water as well as evaporation to the soap paste. Maintain a slow, steady boil throughout the cook. Check the soap paste for separation every 20 minutes or so during the first hour, being careful when removing the lid to avoid burns from the escaping steam. Stir the paste and continue until the soap achieves translucency.

Crock Pot or Turkey Roaster Process: The crock pot or turkey roaster process is very popular with soapmakers. The main concern in this process is having a crock pot with a setting low enough to avoid boil-overs. The majority of them will work very nicely. If in doubt, start with a test recipe low in coconut oil. Be advised once the crock pot is used for soapmaking, it should only be used for soapmaking.

Start the process from the beginning of the directions by melting the oils in the crock pot on low setting. Once the oils have melted, follow the soapmaking instructions as outlined above until full trace is achieved. Place the lid on the crock pot and stir every thirty minutes or so making sure to scrape down the sides and bottom to prevent scorching. Cook on low or warm depending on the heat of the individual crockpot until the soap achieves translucency.

A turkey roaster will work for batches too large for the crockpot and the thermostat will allow for better temperature control. The directions are the same as using the crock pot.

Another method using the turkey roaster is the water bath process. This method allows for more than one recipe to cook at a time and also eliminates scorching, evaporation and the need for stirring. This is a good method to use in testing recipes.

Water Bath Process: The water bath process is much like the double boiler process. Three to four inches (8–10 cm) of water is placed in the turkey roaster and heated to just below boiling 180°–195°F (82°–91°C). The soap is transferred into crock pot liners, boiling bags or freezer bags and securely tied. They are then placed on a rack into the steaming water with the ties above the water line. Allow to cook until translucency.

Oven Process: The oven process can be done one of two ways.

1. Preheat the oven between 180°–195°F (82°–91°C). Place the cooking pot in the oven with a well-fitting lid (aluminum foil may be placed over the pot as long as it does not come in contact with the raw soap). Place a pan of boiling water in the oven to provide moisture and prevent excess evaporation. Cook the soap at the required temperature, checking and stirring every half hour or so until the soap reaches translucency.

2. This second method eliminates much of the danger of overheating and evaporation associated with the first method. It also uses less energy.

    Depending on the recipe, preheat the oven between 180°–195°F (82°–91°C) Place the cooking pot in the oven with a well-fitting lid (aluminum foil may be placed over the pot as long as it does not come in contact with the raw soap). Cook for one hour checking half-way through to ensure the soap has not separated. Turn off the oven and allow the cook pot to remain in the oven for at least eight hours or overnight if preparing the soap in the evening. Do not open the oven door during the entire process. Depending on the insulating ability of the oven, the soap should achieve translucency during this time.

    Some ovens retain heat better than others. The oven in my soap shop was the least expensive model on the showroom floor fifteen years ago. Needless to say the insulation value is very poor. Whenever I want to make soap in this manner I use a method similar to the oven process but it eliminates the need for any external heat source.

No Cook Process: Line a heavy styrofoam or other well insulated cooler with towels or a blanket. As soon as the soap is ready for the cook, place the soap pot in the cooler and cover completely with the towels or blanket for further insulation. Shut the lid tightly and allow the soap pot to remain in the cooler for at least twelve hours. Additional insulation may be used by placing a blanket over the cooler. The soap should achieve translucency during this time.

The no cook process does have a downside. It is important the paste is at its optimum temperature when placed in the insulated cooler [183°–190°F (84°–88°C)]. It works best on batches of 5 pounds (2.3 Kg) of oil or larger. Smaller batches have difficulty maintaining temperature. It also works best on recipes higher in coconut oil. Soft oil recipes take much longer to fully saponify.

Step 11 Check for Neutrality

Once the soap has achieved translucency, it is time to check for neutrality. As has been discussed earlier, neutrality in soap means having no free alkali or fatty acids. Checking the soap for unsaponified oils is simple and only requires dilution. Checking the soap for free alkali requires phenolphthalein, a pH indicator in solution. Since we will be working in smaller quantities, we will be weighing in grams for better accuracy.

1. **Checking for unsaponified oils:** Dissolve a small sample of soap in about twice the amount of water. The weight or rate of dilution does not have to be precise at this time. Once the soap solution has dissolved check the clarity of the solution.

   - If it is clear, the fatty acids have bound with the lye and you are now ready to check for excess alkali.

   - If the soap solution is cloudy, there are still unsaponified oils present in the soap. This is most likely due either to inadequate cook time or inadequate lye in the recipe. The test for excess alkalinity will indicate which measure to address.

2. **Checking for free alkali:** Place a few drops of phenolphthalein in the diluted soap solution and gently swirl, making sure no lather is present in the solution. Lather will indicate the presence of excess alkali even if the soap is neutral. This is due to a very small amount of the soap in the lather breaking down into free fatty acid and free caustic due to hydrolysis (the decomposition of a chemical compound by its reaction with water). If lather is a problem in the test solution, spritz a small amount of denatured or ethyl alcohol in the solution to dampen it. The results of the phenolphthalein test will indicate which steps will be necessary to achieve a neutral soap.

   - If the soap solution is cloudy but shows no color change with the addition of the phenolphthalein, the soap has unsaponified oils and the lye will need to be adjusted. See the instructions on correcting excess oils in "Step 13: Neutralizing the Soap" before proceeding any further.

   - If the solution is cloudy and also shows pink, a longer cook time is indicated. Place the soap pot back on the heat and cook for another hour. Take another sample and repeat steps one and two. If the new sample continues to remain cloudy and pink, continue cooking for up to an additional two hours. If the problem persists, it could either be the water or the oils. Hard water will make cloudy soap. Oils and waxes with a large percentage of unsaponifiables or saturated fats will cloud the soap as well. The soap will need to be sequestered, which is discussed in "Step 14: Settling the Soap."

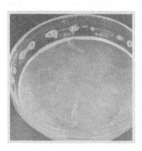

- If the soap solution is clear and pink, this indicates excess alkali and the soap will need to be neutralized. The color can range from pastel to hot pink. The darker the shade, the more neutralizer will be necessary. Follow one of the methods outlined in "Step 13: Neutralizing Excess Alkali."

- If the solution is clear and there is no color change, no neutralization is necessary and the soap paste is ready to dilute.

Before proceeding to the next step, allow the soap to settle for the next several days. Experience has shown, even after the soap tests clear, the fatty acids and excess alkali will continue to unite for several days after the process has been declared finished. Finishing the soap too soon risks over-neutralizing, resulting in sedimentation of free fatty acids to the bottom of the container and a higher risk of oxidation to the soap.

Step 12 Diluting the Soap Paste

The soap paste recipes in this book are made with approximately 33% water and the rest is what is understood as anhydrous soap (soap plus alkalis). Depending on the oils, some soap pastes may be made with up to 60% water. Pastes with a high percentage of coconut oil can absorb more water than pastes made with soft oils. If too much water is added, the excess will weep out of the paste.

The reverse can be said for diluted soap stock. Too high a percentage of anhydrous soap in the stock will cause a portion of the soap to revert back to paste in the form of a gel floating on top of the lovely clear soap. The thicker the layer, the more water will be required to dissolve it. The minimum dilution rate varies depending on the oils in the soap stock.

Some oils will allow a higher concentration of soap-to-water than others. Castor oil will allow a concentration of up to 48% anhydrous soap, coconut oil 41% and most of the soft oils require a concentration no higher than 25–30%. A blend of oils including coconut or castor will allow a more concentrated soap solution than if using only soft oils. Most liquid soaps are diluted in the 25% range with higher concentrations being considered concentrated. The table on page 33 will help achieve the desired concentration based on the oils in the recipe.

To calculate the amount of additional water required to achieve the desired concentration of anhydrous soap in the product,

1. Weigh the soap paste to be diluted and record the weight. If planning on diluting the entire amount of paste, it will be beneficial to record the weight of the empty soap pot before getting started.

> Weighing the empty pot before getting started has always been my most forgotten step. I finally broke down and numbered, weighed and recorded the weight of all my pots to avoid future frustration.

2. Determine the desired concentration of soap and multiply the amount of soap paste being diluted by the corresponding number in the adjacent chart. This is the amount of water; including any other liquid additives (such as glycerin, essential oils, fragrance and neutralizer) necessary to dilute the soap paste.

Percentage of Soap in Solution	Multiply weight of Paste by
40%	0.67
35%	0.91
30%	1.22
25%	1.67
20%	2.33
15%	3.45

Example Most of the soap paste recipes in this book make a total of 67–70 oz. of soap paste. To dilute the whole recipe, multiply the total amount of the paste given in the recipe by the number in the column corresponding to the percentage of anhydrous soap you desire in the finished soap: This gives the amount of water and/or other liquids needed to dilute the soap paste to the desired percentage. Looking at the chart above and assuming the soap paste weighed a total of 70 oz. (1,985 g):

For a 40% anhydrous soap: Multiply 70 oz. (1,985 g) paste × 0.67 = 46.9 oz. (1,330 g) water necessary to dilute the soap paste.

For a 35% anhydrous soap: Multiply 70 oz. (1,985 g) paste × 0.91 = 63.7 oz. (1,806 g) additional liquid.

For a 30% anhydrous soap: Multiply 70 oz. (1,985 g) paste × 1.22 = 85.4 oz. (2,421 g) additional liquid.

Dilution of the Paste

Weigh out the required amount of distilled or soft water according to the dilution chart.

- Bring the water to a boil and add the soap paste.
- Turn down the heat and simmer until the soap paste is warmed through.
- Once the soap becomes soft and pliant, take a stick blender and blend the paste into the water.
- Turn off the heat.
- Cover the soap pot and allow the foam to subside and the rest of the paste to dissolve.
- Once the paste has dissolved, the soap is ready to be finished. Neutralize using one of the following methods outlined in Step 13.

Step 13 Neutralize the Soap

"Free Alkali or Acid (Method for Liquid Soaps).—Dissolve 10 grams of the sample in about 200 cc of freshly boiled, neutral absolute alcohol. Filter and wash the precipitate with freshly boiled, neutral alcohol. Unite the filtrate and washings, add phenolphthalein, and titrate with standard acid or alkali. Calculate the percentage of free alkali as KOH or free acid as oleic acid." Specifications for and Methods of Testing Soaps, issued December 16, 1916 (United States, National Bureau of Standards.)

There are several different ways soap can be neutralized ranging from the very scientific to the approximate. For the very scientific I would highly recommend the book *Scientific Soapmaking* (Dunn, 2010).

A Fairly Accurate Method of Determining the Amount of Neutralizer Required

The following directions are very loosely based on the procedures outlined in *Scientific Soapmaking* (Dunn 2010). Don't let the directions scare you. It's not as laborious or as complicated as it first appears. It only takes a few minutes once you've gotten the concept and it will be time well spent!

All weights will be in grams for better accuracy.

To neutralize excess alkali using this method, you will need the following:

- Citric acid
- Phenolphthalein
- Denatured alcohol or distilled water
- Pipets or eye droppers
- Disposable paper or plastic cups 4 ounces (113 grams) or larger
- Soap scale
- Centigram balance (weighing scale that reads to 0.01 grams), a small investment of money well spent with a host of other uses in the soap shop.
- Calculator
- Pen and paper
- Gloves and goggles

1. **Prepare the neutralizer:** Prepare a 20% citric acid solution by dissolving 28 grams citric acid in 113 grams boiling distilled water. It is not necessary to be absolutely precise as long as this is the solution that will be neutralizing the entire batch of soap; otherwise it is extremely important the numbers be recorded so the solution can be replicated.

2. **Dissolve a sample of the soap paste in either denatured alcohol or water:** Alcohol soap solutions reach the endpoint of a phenolphthalein titration more decisively than water and soap solutions, giving more reliable results. The alcohol method is the preferred choice, but adequate results can also be achieved with soap paste diluted in water.

   Since alcohol is more acidic than soap, it is necessary to adjust the pH of the alcohol to approximately the same pH as soap. This is extremely quick and easy.

   a. Pour 8 ounces (227 grams) of denatured alcohol into a glass jar or plastic bottle. Denatured alcohol, found in the paint department at local hardware stores, is acceptable for testing purposes but should not be used for making soap or cosmetics.

   b. Place a few drops of phenolphthalein into the alcohol.

   c. Place a flake of potassium hydroxide into the alcohol and stir to dissolve. The alcohol solution should show a very slight hint of pink, if not, continue one flake at a time until a color change is noted.

Dissolve approximately 60 grams of soap paste in 120 grams of either the pH adjusted denatured alcohol or water. This will be your test sample.

3. **Prepare the testing samples:** Label 3 disposable cups A, B and C. Place cup A on the centigram balance and tare the weight. Add 50 +/- grams of diluted soap stock. Again, these weights do not have to be exact but they do need to be recorded in your notes. Repeat this step with cups B and C recording the weight of each.

   Example If the weight of cup A was 50.44, Cup B was 50.99 and Cup C was 50.40, you would write down each number beside the corresponding letter of the cup.

   Soap already diluted including No Paste and Gel Soap will give more reliable test results with the addition of approximately 10 grams neutralized alcohol weighed into each soap sample. Do not include the weight of the alcohol in the calculations as its only benefit is for the reaction of the phenolphthalein.

   If using a stir stick to help incorporate the citric acid, place the stir stick in the cup and tare before adding the citric acid so its weight is not included. Keep the stir stick in the cup until the final weight of the citric acid has been recorded.

4. **Neutralize the three samples:** Place a few drops of phenolphthalein in each cup. Place cup A on the centigram balance and tare. Using a pipet, add the citric acid solution to the soap solution a few drops at a time. Swirl gently between

each addition until the solution begins to change color. Once a color change is noticeable slowly add the citric acid solution by only one drop at a time until it no longer shows pink. Record the weight of the citric acid solution. Repeat with cups B and C.

Example If the weight of the citric acid solution added in cup A was 0.29 grams, cup B was 0.30 grams and cup C was 0.35 you would record these numbers beside the corresponding letter.

5. **Determine the percentage of citric acid required to neutralize the soap:** Divide the number of grams of citric acid solution required for cup A by the number of grams of soap in cup A to get the percentage of citric acid solution required to neutralize the soap. Repeat for B and C.

Example If cup A required 0.29 grams of citric acid solution to neutralize 50.44 grams of soap:
$0.29 \div 50.44 = 0.0057$ or 0.6% citric acid solution required to neutralize the soap.

Cup B required 0.30 grams of citric acid solution to neutralize the 50.99 grams of soap solution: $0.30 \div 50.99 = 0.0058$ or 0.6% citric acid solution required to neutralize the soap.

Cup C required 0.35 grams of citric acid solution to neutralize 50.40 grams of soap solution: 0.36 ÷ 50.40 = 0.0069 or 0.7% citric acid solution required to neutralize the soap.

Compare the results of the three samples: Compare the answers of cups A, B and C. If at least two of the three numbers are close in value, take the average of the two numbers. This is the percentage of citric acid solution you will use to neutralize the soap. In the example, cup A and cup B both required 0.6% citric acid solution to neutralize the soap. We will use that percentage to calculate how much citric acid solution is required to neutralize by multiplying 0.006 time the weight of the diluted soap stock.

Recommendation: Without getting into analytical and synthetic weighing (explained so well by Kevin Dunn and beyond the scope of this book) as well as the pH of varying concentrations of dilution, it is impossible to be completely accurate in these calculations. These are ballpark figures to get the soap very close to neutral.

Start adding the citric acid solution to the soap. It would be wise to measure out the required citric acid solution and add ¾ or less of the total solution (depending on how confident you are with your titrating and weighing skills). Before adding any additional neutralizer, take a quick phenolphthalein test. The test results will indicate how close you are to the endpoint. If there is very little pink, proceed slowly. If there is no pink, aren't you glad you stopped when you did?

Once you become more familiar with the neutralizing requirements of your particular recipes and with your particular raw ingredients, you can proceed with more confidence.

An Approximate Method of Neutralizing Soap Stock with Excess Alkali

This is the procedure most handcrafted soapmakers employ. The directions are for diluting an entire batch of soap. If diluting smaller quantities use smaller measuring devices such as teaspoons and pipets or eye droppers in the place of the tablespoons and teaspoons called for in these directions. It requires patience and care to avoid over neutralizing. You will need the following equipment and supplies when utilizing this method:

- Citric acid
- Distilled or soft water

- Tablespoons and teaspoons
- Soap scale
- Disposable paper or plastic cups or small jars
- Phenolphthalein solution
- Eye dropper or pipet
- Pen and paper
- Gloves and goggles

1. **Prepare the neutralizer:** Prepare a 20% citric acid solution by dissolving 28 grams citric acid in 113 grams boiling distilled water.

2. **Dilute the soap:** Dilute the soap according to the directions in Step 12.

3. **Neutralize the diluted soap.** Add a tablespoon at a time of citric acid solution to the warm soap and incorporate fully. Test a sample of the solution with phenolphthalein. Repeat until there is a noticeable change in the shade of pink. Once the color is a very pale pink, add the neutralizer one teaspoon at a time until no pink remains.

Although the above method sounds very simple and straightforward, experience will convince you appearances can be deceiving. It isn't nearly as uncomplicated as it seems and requires more time than it appears.

Neutralizing the Soap Stock with Stearic Acid

This easy, almost fail-proof method has many advantages as well as a few drawbacks. The soap must first be in a liquid state and works best when using the "No Paste Liquid Soapmaking Method" outlined in Chapter 5. To avoid clouding and over-neutralizing, make sure the soap is over-concentrated. This will be evident by the formation of a thin filmy layer on top of the soap. Using stearic acid to neutralize liquid soap is like having a Pac-Man forager swirling around the soap pot and latching on to all the free alkali. Once the alkali has been vanquished, the excess stearic acid puffs up and floats to the top of the soap pot to be strained away. The film will capture the stearic acid and prevent it from dissolving into the soap.

To neutralize excess alkali with stearic acid, you will need the following equipment and supplies:

- Chelating agent (An optional ingredient that helps bind minerals in the soap and is discussed more fully in "Chapter 13: Shelf Life and Stability").
- Phenolphthalein
- Stearic acid
- Pot or glass measuring cup to melt the stearic acid
- Pen and paper
- Goggles and gloves
- Bowl or additional soap pot
- Colander or strainer
- Cheesecloth, paper towel or clean lint free cloth for filtering

> Unlike citric and gluconic acids, stearic acid does not contribute chelates to the liquid soap. Add any chelating agents prior to neutralizing with stearic acid to avoid over-neutralizing with the addition of a chelate when the soap is already completely neutral.

1. **Estimate the amount of stearic acid required:** Depending on the intensity of color in the phenolphthalein test, calculate 0.1–0.3 ounces (3–9 grams) or 1–2 teaspoons of stearic acid per pound of oil. A pale pink will indicate only a small amount of stearic acid is required while a darker pink will indicate the addition of more stearic acid.

2. **Melt the stearic acid** in a pot on the stove or in a microwave safe container in the microwave.

3. **Bring the soap to temperature:** It is important the soap solution is warm enough to incorporate the stearic acid. The temperature should be between 130°–150°F (54°–66°C).

4. **Add the stearic acid to the hot soap:** Stir the stearic acid thoroughly into the hot soap making sure it has the opportunity to travel to all parts of the soap pot. Not all of the stearic acid will completely dissolve into the soap which is what you want. The excess stearic acid assures the soap will be neutral.

5. **Put a lid on it:** Cover the soap and allow it to cool to room temperature.

6. **Strain the excess stearic acid:** Once the solution has cooled, the excess stearic acid will precipitate out of the soap solution and float to the top in the form of a white curd. Skim off as much as possible before straining.

    - Place a strainer or colander over a bowl or soap pot. Allow enough clearance that the strained soap avoids coming in contact with the bottom of the strainer.

    - Place a paper towel, cheesecloth or clean lint free cloth in the strainer or colander to catch the smaller particles.

    - Pour the liquid soap into the strainer and allow it to drain into the bowl or soap pot being careful not to overfill the strainer.

    - Just to be on the safe side, take a small amount of the soap and check for neutrality.

> Using stearic acid as a neutralizer with formulas already high in saturated fatty acids or waxes may cloud the soap.

Neutralizing Excess Oils

If the soap sample is cloudy and phenolphthalein shows no color change, the soap has excess oils and additional alkali is necessary to create a neutral soap. Hitting the neutral point without going over when neutralizing excess oils is the sticky wicket that makes neutralizing excess alkali the more desirable choice. With a little more caution and time, neutralizing excess oils is achievable.

- Instead of creating an acid solution, create a strong potassium hydroxide solution by dissolving 2 ounces (57 grams) potassium hydroxide in 3 ounces (85 grams) of water.

- Dilute the soap paste to a 40% concentration using the directions in "Step 12: Diluting the Soap Paste." It is much easier to correct a highly concentrated soap stock than it is to correct a highly diluted one.

- Incorporate 1–2 ounces (28–57 grams) of the potassium hydroxide solution to the diluted soap stock and cook for one hour with occasional stirring.

- At the end of the cook time, remove a small sample of the soap stock and place a few drops of phenolphthalein into the test sample. If there is no color change, continue by adding another ounce of the lye solution to the soap stock and cook for an additional hour. Continue adding the lye solution until the soap becomes clear and a slight color change is noted. Allow the soap to settle for a couple of days. At this time any excess alkali should be taken up with the fatty acids. Test again and if the soap solution continues to show pink, neutralize according to the soapmaker's choice of neutralizing excess alkali.

Step 14 Settling or Sequestering the Soap

Once the soap has been neutralized it's a good idea to allow the soap to settle for a week or so. This allows any unsaponifiables and free fatty acids to sink to the bottom, greatly reducing cloudiness and providing additional sparkle to the finished product. This process can be accelerated by placing the soap in the refrigerator for 24–48 hours. After refrigeration, allow the soap to continue settling for a couple of days. Carefully draw off the clear soap leaving the residue to be recycled into laundry or dish soap. At this point the soap can be used as is or it can be thickened (Chapter 7), fragranced (Chapter 11) and/or colored (Chapter 10).

CHAPTER

SOAP PASTE RECIPES

"Liquid Glycerine Soap. Melt together 374 lb. pale oleic acid, 66 lb. coconut oil, 228 lb. caustic potash lye, 60°TW., then add, boil up, and when saponified add 20 lb. glycerine and enough methylated spirit to make the liquid clear."

~ Soaps (HURST, 1898)

The recipes in this section are basic recipes that can be enhanced and personalized with colorants, thickeners and other additives to achieve impressive soaps suitable for most applications. Refer to the dilution chart in Appendix B to dilute to the desired concentration.

Most of the recipes are calculated using coconut and olive oil as the majority oils in the formula. Palm kernel oil may be substituted for the coconut oil, but the lye portion of the recipe will need to be recalculated. Soft oils including high oleic sunflower, high oleic safflower, apricot kernel, avocado and canola can be substituted for the olive oil without adjustments to the lye calculations. Sweet almond, rice bran, hemp, macadamia nut, meadowfoam seed and other oils whose SAP values differ by more than 2 points from the average saponification value of olive oil should not be substituted unless the lye portion of the recipe is recalculated. Refer to "Chapter 15: How to Formulate Liquid Soap Recipes" if recalculating saponification values or making other changes to the formula.

All recipes are presented in ounces with grams in parenthesis.

High Coconut Oil Recipes

High foaming in soft or hard water, coconut oil soaps are very cleansing but can be drying to the skin. Pairing the coconut oil with other oils makes for a better balance of cleaning efficiency and emollience. They make a very stiff paste easily dissolved into a liquid soap the consistency of water. They can be thickened with a cellulose derivative such as hydroxyethylcellulose. Directions can be found on page 75.

> All recipes are calculated to the nearest tenth of an ounce or to the nearest gram. While it is not necessary to be this precise in the weighing of the ingredients, having precise values will allow the soapmaker to double or triple a recipe in confidence.

100% Coconut Oil Soap

Reserve this soap for tough jobs and oily complexions! For dry and sensitive skin, use the recipes with added soft oils to lend nourishing moisture to the soap.

Oils:

36 oz. (1,021 g) Coconut Oil

Lye Solution:

10.1 oz. (286 g) Potassium Hydroxide
24 oz. (680g) Water

Total Weight of Paste = 70.1 oz. (1,987 g)

80% Coconut Oil Soap w/ Olive Oil

Oil:

28.8 oz. (817 g) Coconut Oil
7.2 oz. (204 g) Olive Oil

Lye Solution:

9.6 oz. (272 g) Potassium Hydroxide
23.5 oz. (666 g) Water

Total Weight of Paste = 69.1 oz. (1,960 g)

75% Coconut Oil Soap w/ Olive Oil

Oils:

27 oz. (765 g) Coconut Oil
9 oz. (255 g) Olive Oil

Lye Solution:

9.5 oz. (269 g) Potassium Hydroxide
23.5 oz. (666 g) Water

Total Weight of Soap Paste = 69 oz. (1,956 g)

75% Coconut Oil Soap w/ Olive & Castor Oils

Castor oil adds cleansing and moisturizing properties and contributes transparency to the soap.

Oils:

27 oz. (765 g) Coconut Oil
7.2 oz. (204 g) Olive Oil
1.8 oz. (51 g) Castor Oil

Lye Solution:

9.5 oz. (269 g) Potassium Hydroxide
23.5 oz. (666 g) Water

Total Weight of Soap Paste = 69 oz. (1,956 g)

75% Coconut Oil Soap
w/ Castor Oil & Cocoa Butter

Cocoa butter gives the soap body and creamy lasting bubbles while nourishing the skin!

Oils:

27 oz. (765 g) Coconut Oil

5.4 oz. (153 g) Castor Oil

3.6 oz. (102 g) Cocoa Butter

Lye Solution:

9.4 oz. (266 g) Potassium Hydroxide

23.4 oz. (663 g) Water

Total Weight of Soap Paste = 68.8 oz. (1,950 g)

68% Coconut Oil Soap
w/ Rice Bran and Castor Oils

Castor oil helps clarify cloudiness created by the rice bran oil

Oils:

24.5 oz. (695 g) Coconut Oil

7.2 oz. (204 g) Rice Bran Oil

4.3 oz. (122 g) Castor Oil

Lye Solution:

9.2 oz. (261 g) Potassium Hydroxide

23.3 oz. (661 g) Water

Total Weight of Soap Paste = 68.5 oz. (1,942 g)

60% COCONUT OIL SOAP w/ OLIVE & CASTOR OILS

A nice balance of lather and emollience.

Oils:

21.6 oz. (612 g) Coconut Oil

10.8 oz. (306 g) Olive Oil

3.6 oz. (102 g) Castor Oil

Lye Solution:

9 oz. (255 g) Potassium Hydroxide

23 oz. (652 g) Water

Total Weight of Soap Paste = 68 oz. (1,928 g)

60% COCONUT OIL SOAP w/ OLIVE, CASTOR & PALM OILS

Palm oil helps stabilize the lather and gives body to the soap.

Oils:

21.6 oz. (612 g) Coconut Oil

7.2 oz. (204 g) Olive Oil

3.6 oz. (102 g) Palm Oil

3.6 oz. (102 g) Castor Oil

Lye Solution:

9.1 oz. (258 g) Potassium Hydroxide

23 oz. (652 g) Water

Total Weight of Soap Paste = 68.1 oz. (1,931 g)

60% Coconut Oil Soap w/ Mango Butter, Avocado and Castor Oils

Designed for mature skin, mango butter, avocado and castor oil add rejuvenating, skin healing properties to the soap.

Oils:

21.6 oz. (612 g) Coconut Oil

7.2 oz. (204 g) Avocado Oil

5.4 oz. (153 g) Castor Oil

1.8 oz. (51 g) Mango Butter

Lye Solution:

9 oz. (255 g) Potassium Hydroxide

23 oz. (652 g) Water

Total Weight of Soap Paste = 68 oz. (1,928 g)

58% Coconut Oil Soap w/ Palm, Macadamia Nut & Castor Oils

Macadamia and palm oil have excellent antioxidant properties.

Oils:

20.8 oz. (589 g) Coconut Oil

5.8 oz. (164 g) Macadamia Nut Oil

5.4 oz. (153 g) Palm Oil

4 oz. (113 g) Castor Oil

Lye Solution:

9.1 oz. (258 g) Potassium Hydroxide

23 oz. (652 g) Water

Total Weight of Soap Paste = 68.1 oz. (1,931 g)

50% Coconut Oil Soap w/ Olive Oil

Oils:

18 oz. (510 g) Coconut Oil

18 oz. (510 g) Olive Oil

Lye Solution:

8.8 oz. (249 g) Potassium Hydroxide

23 oz. (652 g) Water

Total Weight of Paste = 67.8 oz. (1,922 g)

High Soft Oil Recipes

Soaps high in soft oils such as olive and high oleic sunflower oil are extremely stubborn to get to trace, require a longer cook time, have a close lather and are more difficult to dissolve than their coconut oil soap counterpart. What makes them an excellent choice in liquid soapmaking is their gentleness on the skin, their higher viscosity and the ease in which they may be naturally thickened with common salt. Castile soap is one such example. Touted by kings and noblemen since the 11th century, it continues to be regarded as the finest soap on the market today. Historically, castile soap was made from the ashes of the barilla tree, olive oil and laurel fruit oil. The laurel oil was dropped when price and availability made it too difficult to obtain and sodium hydroxide replaced the ashes. Today soap may be considered Castile, whether liquid or bar soap, but the controversy continues in regard to whether a soap may be considered true Castile if a percentage of the oils in the formula are something other than olive oil.

Because of the slow nature of achieving trace in soaps with a high percentage of soft oils, a couple of tips may help cut down on the amount of quality time spent with the stick blender:

- Mix the oils and lye together, turn off the heat and allow the mixture to stand for several hours or overnight before bringing to trace. The slight saponification of the oils during this time will help bring the rest of the oils and lye into a quicker emulsion. Bring to temperature and proceed as usual.

🦆 The addition of accelerants to the formula, such as a small amount of soap from a previous recipe and/or glycerin will speed things along (For more on accelerants, refer to page 60). Add just after mixing the oils and lye together and proceed as usual.

85% Olive Oil Soap w/ Coconut Oil

Oils:

30.6 oz. (868 g) Olive Oil

5.4 oz. (868 g) Coconut Oil

Lye Solution:

7.8 oz. (221 g) Potassium Hydroxide

22.5 oz. (638 g) Water

Total Weight of Soap Paste = 66.3 oz. (1,880 g)

80% Olive Oil Soap w/ Coconut and Castor Oils

Oils:

28.8 oz. (817 g) Olive Oil

5.4 oz. (153 g) Coconut Oil

1.8 oz. (51 g) Castor Oil

Lye Solution:

7.8 oz. (221 g) Potassium Hydroxide

23 oz. (652 g) Water

Total Weight of Soap Paste = 66.8 oz. (1,894 g)

75% High Oleic Sunflower Oil Soap w/ Coconut Oil, Castor Oil and Cocoa Butter

Oils:

27 oz. (765 g) High Oleic Sunflower Oil

1.8 oz. (51 g) Cocoa Butter

3.6 oz. (102 g) Coconut Oil

3.6 oz. (102 g) Castor Oil

Lye Solution:

7.8 oz. (221 g) Potassium Hydroxide

23 oz. (652 g) Water

Total Weight of Soap Paste = 66.8 oz. (1,894 g)

90% Olive Oil Soap w/ Coconut Oil

Oils:

32.4 oz. (919 g) Olive Oil

3.6 oz. (102 g) Coconut Oil

Lye Solution:

7.7 oz. (218 g) Potassium Hydroxide

22.5 oz. (638 g) Water

Total Weight of Soap Paste = 66.2 oz. (1,877 g)

75% Olive Oil Soap
w/ Avocado and Castor Oils

Oils:

27 oz. (765 g) Olive Oil

5.4 oz. (153 g) Avocado Oil

3.6 oz. (102 g) Castor Oil

Lye Solution:

7.4 oz. (210 g) Potassium Hydroxide

22 oz. (624 g) Water

Total Weight of Soap Paste = 65.4 oz. (1,854 g)

85% High Oleic Sunflower
Oil Soap w/ Sweet Almond Oil

Oils:

30.6 oz. (868 g) High Oleic Sunflower Oil

5.4 oz. (153 g) Sweet Almond Oil

Lye Solution:

7.6 oz. (216 g) Potassium Hydroxide

22.5 oz. (638 g) Water

Total Weight of Soap Paste = 66 oz. (1,874 g)

90% Olive Oil Soap w/ Castor Oil

Oils:

32.4 oz. (919 g) Olive Oil
3.6 oz. (102 g) Castor Oil

Lye Solution:

7.4 oz. (210 g) Potassium Hydroxide
22 oz. (624 g) Water

Total Weight of Soap Paste = 65 oz. (1,854 g)

75% Olive Oil Soap w/ Palm, Coconut And Castor Oils

Oils:

27 oz. (765 g) Olive Oil
1.8 oz. (51 g) Palm Oil
3.6 oz. (105 g) Coconut Oil
3.6 oz. (105 g) Castor Oil

Lye Solution:

7.7 oz. (218 g) Potassium Hydroxide
22.5 oz. (638 g) Water

Total Weight of Soap Paste = 66.2oz. (1,877 g)

100% Olive Oil Soap

Oils:

36 oz. (1,021 g) Olive Oil

Lye Solution:

7.5 oz. (213 g) Potassium Hydroxide
22.5 oz. (638 g) Water

Total Weight of Soap Paste = 66 oz. (1,871 g)

65% Olive Oil Soap
w/ Castor & Coconut Oils

Oils:

23.4 oz. (664 g) Olive Oil
9 oz. (255 g) Coconut Oil
3.6 oz. (102 g) Castor Oil

Lye Solution:

8.1 oz. (227 g) Potassium Hydroxide
23 oz. (652 g) Water

Total Weight of Soap Paste = 67.1 oz. (1,902 g)

70% OLIVE OIL SOAP
w/ COCONUT OIL

Oils:

25.2 oz. (714 g) Olive Oil

10.8 oz. (306 g) Coconut Oil

Lye Solution:

8.2 oz. (233 g) Potassium Hydroxide

23 oz. (652 g) Water

Total Weight of Soap Paste = 67 oz. (1,905 g)

75% OLIVE OIL SOAP
w/ COCONUT OIL

Oils:

27 oz. (765 g) Olive Oil

9 oz. (255 g) Coconut Oil

Lye Solution:

8.1 oz. (229 g) Potassium Hydroxide

23 oz. (652 g) Water

Total Weight of Soap Paste = 67.1 oz. (1,902 g)

CHAPTER

HOW TO MAKE LIQUID SOAP USING THE NO PASTE METHOD

No Paste Soapmaking Method

Large scale liquid soap production is carried out in large vats with closed steam coils equipped with agitators or mixers. The oil is introduced, heat is applied and the lye is run through with sufficient water to obtain a 35% soap solution. A constant temperature is maintained over a period of many hours while the solution slowly agitates. At complete saponification, samples are drawn and tested for free fatty acids or free alkali and adjustments are made. Additional water is added, or the soap is cooked down to bring it to the desired concentration. The soap is allowed to settle for a few weeks. Some soapmakers employ chillers in which the soap is chilled to below 40°F (4.44°C) for 24 to 48 hours and filtered. It is then returned to room temperature for a week or more and allowed to settle. The clear soap is drawn off and the sediment is cleaned out.

∽ Summarized From: *Modern Soapmaking*
(THOMSSEN AND KEMP, 1937)

Sanitation and janitorial supply industries as well as larger natural soap companies, such as Dr. Bronner's and Vermont Soap, continue to produce true liquid soap using the same or similar methods to those just mentioned above. The handcrafted soapmaker could greatly benefit from the more streamlined method of making liquid soap without the necessity of first creating a paste. The main obstacle is the lack of safe and affordable equipment in handcrafted sizes. By eliminating the need for the costly equipment with substitutions of ingredients to simulate their actions, the handcrafted soapmaker can make beautiful liquid soaps without pastes. The secret is in the choice of oils, the use of solvents and accelerants and a combination of alkalis including potassium carbonate.

The process is surprisingly fast and easy with only a few variations from the paste method. Experience with making paste soaps before tackling the no paste method is strongly recommended.

Coconut Oil

This method needs to move along quickly and coconut oil is willing to oblige. Its extreme solubility and affinity for the soap pot is unparalleled. Other oils may be included in the formula, but the majority of the oil in the formula should be coconut oil or another oil high in lauric acid such as palm kernel or babassu oil. Adjustments in the lye calculations will be necessary to offset the difference in the SAP values if substitutions are made.

Potassium Hydroxide and Sodium Hydroxide

Using a percentage of sodium hydroxide with the required potassium hydroxide will create a synergistic effect, extending the energy of the combined alkalis. This is especially helpful when making soaps with additives or butters or just to hurry things along. It is an optional ingredient that will also add body and help stabilize the lather.

Potassium Carbonate and Glycerin

A liquid soap high in coconut oil can be achieved without the need for mechanical mixing with the addition of either glycerin or potassium carbonate. Both are not necessarily warranted, but the results are quicker and the amount of dilution and cook time is less when both are added. Soaps in higher concentration saponify more quickly. Creating the highest concentration of soap stock that maintains a liquid state is essential to this process. Glycerin, with its solvent action allows the soap to stay fluid in higher soap concentrations. It also accelerates saponification and acts as a clarifier. Other solvents such as specially denatured ethyl alcohol, sorbitol or propylene glycol, which are more efficient as solvents and accelerants, may be substituted for the glycerin. The choice of glycerin is mainly due to consumer appeal and the fact most liquid soapmakers already have it in their inventory.

Potassium carbonate, while not a solvent, plays a similar role in this soapmaking process. It eliminates continual mixing by placing itself between the soap molecules allowing the soap to stay fluid in higher concentrations.

Instructions for the no paste soapmaking method are very similar to the paste method. For brevity and continuity, the directions exactly the same as making a soap paste are briefly outlined and more detailed information is given to instructions differing from the paste directions beginning on page 23.

No Paste Soapmaking Instructions

Step 1 **Safety First!** Goggles, gloves protective mask and clothing!

Step 2 **Gather Thoughts and Materials:** Make sure you have everything you need.

Step 3 **Weigh and Melt the Oils:** Ideal temperature should be between 160°F–175°F (71°–79°C).

Step 4 **Prepare the Dilution Water and the Potassium Carbonate:** Weigh the potassium carbonate, the dilution water and any other additives called for in the recipe and heat to boiling. Remove from heat and set aside. This will be added at trace.

Step 5 **Prepare the Glycerin:** Weigh the glycerin and heat in the microwave or on the stove until hot but not boiling. Cover and keep warm.

Step 6 **Prepare the Lye Solution:** All of the recipes in this section will include potassium and sodium hydroxide. If the soapmaker prefers to use only potassium hydroxide in the recipe, refer to the soap paste recipe in chapter 4 which corresponds to the recipe number in this chapter for the correct amount of potassium hydroxide required.

The water portion of the lye solution is discounted to allow the addition of glycerin into the lye solution. The glycerin will speed the saponification of the oils. Using it in the lye solution is much more effective than adding it later in the process.

- ❧ Measure the water amount of the lye solution and set aside.

- ❧ Measure the amount of potassium and sodium hydroxide called for in the recipe and add to the water, stirring until dissolved.

🍂 Add the hot glycerin to the dissolved lye solution. Make sure the lye is completely dissolved before adding the hot glycerin. Failure to do so may result in a volcano which could cause serious burns as well as damage to surrounding surfaces!

Step 7 **Stir the Lye Solution into the Oils:** Once the lye solution has been completely dissolved and the glycerin added to the mixture, it should be at a good temperature to add to the oils. Slowly pour the lye solution into the oils and check the temperature. If it is below 160°F (71°C), warm it up until the desired temperature is reached. If it is above 175°F (79°C), allow the oil/lye solution to cool down before stick blending to avoid boil over. Once the desired range of temperature is achieved introduce the stick blender and blend until trace.

Step 8 **Bring to Light Trace:** The definition of trace using the no paste method diverges from the paste method and more closely resembles thick trace in cold process soap, with a few exceptions. Due to the glycerin in the lye solution, the oil/lye mixture will be translucent instead of opaque and creamy. Trace will look more like a gel rather than a pudding (compare thickening sauces with cornstarch rather than flour). A good way to guage trace is when the impression of the stick blender clearly remains in the soap solution when it is removed.

Step 9 **Add the Dilution Water and/or Additives:** Once the soap solution has reached trace, slowly add the dilution water in which the potassium carbonate has been added. Sometimes separation may occur at trace and the trace will have a grainy, applesauce texture. When this happens, add a small amount of the dilution water while stick blending to bring it back into an emulsion. Once the emulsion is creamy and uniform, continue to slowly add the remainder of the dilution water.

Boil over caused by excessive temperature.

Step 10 **Cook the Soap:** The same methods of cooking the soap apply to the no paste method with the addition of being able to cook the soap directly on top of the stove. Caution is advised regarding temperatures. The liquid soap will boil over much more readily than the paste. Maintain a temperature between 180°–190°F (82°–88°C) and periodically stir the soap as it cooks. Add soft or distilled water in 2–3 oz. (57–85 g) increments as necessary to compensate for evaporation. Evaporation will be apparent if a layer of film develops on top of the soap solution or the soap builds up on the spatula or whisk when stirred.

Continue cooking until the soap becomes transparent. Temperature and ingredients will determine the amount of time required. Recipes high in coconut and castor oil will cook much faster (sometimes within an hour) than recipes containing butters or waxes. Some butters do not lend themselves to the no-paste method due to the difficulty of getting them to properly saponify in a reasonable amount of time.

Step 11 Check the Soap for Neutrality.

Step 12 Dilute the Soap to the Desired Concentration.

Step 13 Neutralize the Soap: The amount of water required to achieve the desired concentration of soap is dependent upon the ingredients. In formulating the recipes, consideration has been given to the specific oils in determining the amounts of water required to cook the soaps at their most concentrated levels and still remain liquid. Recipes high in coconut and castor are calculated at 40% anhydrous soap. The soap is ready to use in its concentrated form, or it can be diluted. If using in its concentrated form, make sure the soap is fluid and the bubbles are light and frothy. Scummy bubbles or the formation of a film is a sign the soap is too concentrated and additional water will be required to avoid layering of the soap.

Each recipe will give the total weight of the concentrated soap solution to help in calculating the final dilution rate from the chart below. The dilution chart is approximate. Soap lost through evaporation and stirring will be offset by the addition of water and neutralizer, but not precisely.

No Paste Dilution Chart

Desired Concentration of Soap	Starting @ 40% Soap Multiply by	Starting @ 35% Soap Multiply by	Starting @ 30% Soap Multiply by
35%	0.14	—	—
30%	0.33	0.17	—
25%	0.60	0.40	0.2
20%	1.0	0.75	0.50
15%	1.67	1.33	1.00

Multiply the total weight of the ingredients as listed in the recipe by the number in the corresponding column to get the total amount of liquids necessary to achieve the desired percentage of anhydrous soap in the finished product. Weigh your soap concentrate and compare the total. Adjust the difference by either subtraction or addition, depending on whether the amount in the soap pot is greater (subtract) or lesser (add) than the amount stated in the recipe due to evaporation or over addition of water during the cook.

Step 14 **Allow the Soap to Settle.**

> Adding coconut oil into a potassium soap pot was a revolutionary event in the annals of soapmaking history and the term liquid soap was coined in 1865. Before flip top, pump dispensing closures and plastic squeeze bottles were available to the industry, the water thin consistency of coconut oil soap was considered an advantage in the packaging and distribution of the product. Potassium liquid soaps were easily dispensable in glass bottles and public restroom dispensers. Liquid soap and soft soap maintained their separate identities with little headway into the personal care market due to the problematic packaging of the thicker soft soap and the wastefulness and drying effect of the water thin liquid soap until the introduction of the surfactant Soft Soap™ in 1980 when the revolutionary soap in a pump bottle was introduced.

Step 15 **Finish the Soap:** The liquid soap can now be colored (Chapter 10), thickened (Chapter 7) and scented (Chapter 11).

CHAPTER

NO PASTE RECIPES

"A very good liquid may be prepared by saponifying 5 pounds of cocoanut oil which has been thoroughly purified, with 24 ounces of 98 percent caustic potash, to which is added 2 pounds glycerin, 2 pounds 95 percent alcohol, 3 pounds sugar, 4 or 5 gallons water, the amount of water depending on the consistency of the liquid soap desired."

∽ AMERICAN SOAP MAKER'S GUIDE
(MERBOTT, STANISLAUS, 1928)

100% COCONUT OIL SOAP

Oils:

36 oz. (1,021 g) Coconut Oil

Lye Solution:

7.6 oz. (215 g) Potassium Hydroxide

1.6 oz. (45 g) Sodium Hydroxide

10.6 oz. (301 g) Water

12 oz. (340g) Glycerin

Dilution Water:

45.2 oz. (1,281 g) Water

1.8 oz. (51 g) Potassium Carbonate

Total Weight of 40% Soap Solution = 115 oz. (3,255 g)

80% Coconut Oil Soap w/ Olive Oil

Oils:

    28.8 oz. (817 g) Coconut Oil
    7.2 oz. (204 g) Olive Oil

Lye Solution:

    7.2 oz. (204 g) Potassium Hydroxide
    1.5 oz. (43 g) Sodium Hydroxide
    10.3 oz. (292 g) Water
    12 oz. (340 g) Glycerin

Dilution Water:

    44.7 oz. (1,267 g) Water
    1.8 oz. (51 g) Potassium Carbonate

Total Weight of 40% Soap Solution = 114 oz. (3,218 g)

75% Coconut Oil Soap w/ Olive Oil

Oils:

    27 oz. (765 g) Coconut Oil
    9 oz. (255 g) Olive Oil

Lye Solution:

    7.1 oz. (201 g) Potassium Hydroxide
    1.5 oz. (45 g) Sodium Hydroxide
    10.3 oz. (292 g) Water
    12 oz. (340 g) Glycerin

Dilution Water:

44.6 oz. (1264 g) Water

1.8 oz. (51 g) Potassium Carbonate

Total Weight of 40% Soap Solution = 113.3 oz. (3,212 g)

75% COCONUT OIL SOAP w/ OLIVE & CASTOR OILS

Oils:

27 oz. (765 g) Coconut Oil

7.2 oz. (204 g) Olive Oil

1.8 oz. (51 g) Castor Oil

Lye Solution:

7.1 oz. (201 g) Potassium Hydroxide

1.5 oz. (43 g) Sodium Hydroxide

10.3 oz. (292 g) Water

12 oz. (340 g) Glycerin

Dilution Water:

44.6 oz. (1264 g) Water

1.8 oz. (51 g) Potassium Carbonate

Total Weight of 40% Soap Solution = 113.3 oz. (3,212 g)

75% Coconut Oil Soap
w/ Castor Oil & Cocoa Butter

Oils:

27 oz. (765 g) Coconut Oil

5.4 oz. (153 g) Castor Oil

3.6 oz. (102 g) Cocoa Butter

Lye Solution:

7.1 oz. (201 g) Potassium Hydroxide

1.5 oz. (43 g) Sodium Hydroxide

10.3 oz. (292 g) Water

12 oz. (340 g) Glycerin

Dilution Water:

44.6 oz. (1,264 g) Water

1.8 oz. (51 g) Potassium Carbonate

Total Weight of 40% Soap Solution = 113.3 oz. (3,212 g)

68% Coconut Oil Soap
w/ Rice Bran & Castor Oils

Oils:

24.5 oz. (695 g) Coconut Oil

7.2 oz. (204 g) Rice Bran Oil

4.3 oz. (122 g) Castor Oil

Lye Solution:

6.9 oz. (196 g) Potassium Hydroxide

1.4 oz. (40 g) Sodium Hydroxide

10.1 oz. (286 g) Water

12 oz. (340 g) Glycerin

Dilution Water:

44.4 oz. (1,259 g) Water

1.8 oz. (51 g) Potassium Carbonate

Total Weight of 40% Soap Solution = 112.6 oz. (3,192 g)

60% Coconut Oil Soap w/ Olive & Castor Oils

Oils:

21.6 oz. (612 g) Coconut Oil

10.8 oz. (306 g) Olive Oil

3.6 oz. (102 g) Castor Oil

Lye Solution:

6.8 oz. (198 g) Potassium Hydroxide

1.4 oz. (40 g) Sodium Hydroxide

10 oz. (284 g) Water

12 oz. (340 g) Glycerin

Dilution Water:

44.2 oz. (1,253 g) Water

1.8 oz. (51 g) Potassium Carbonate

Total Weight of 40% Soap Solution = 112.2 oz. (3,181 g)

60% Coconut Oil Soap w/ Olive, Castor & Palm Oils

Oils:

21.6 oz. (612 g) Coconut Oil

7.2 oz. (204 g) Olive Oil

3.6 oz. (102 g) Palm Oil

3.6 oz. (102 g) Castor Oil

Lye Solution:

6.8 oz. (193 g) Potassium Hydroxide

1.5 oz. (43 g) Sodium Hydroxide

10.2 oz. (289 g) Water

12 oz. (340 g) Glycerin

Dilution Water:

44.2 oz. (1,253 g) Water

1.8 oz. (51 g) Potassium Carbonate

Total Weight of 40% Soap Solution = 112.8 oz. (3,198 g)

60% Coconut Oil Soap w/ Mango Butter, Avocado & Castor Oils

Oils:

21.6 oz. (612 g) Coconut Oil

7.2 oz. (204 g) Avocado Oil

5.4 oz. (153 g) Castor Oil

1.8 oz. (51 g) Mango Butter

Lye Solution:

  6.8 oz. (193 g) Potassium Hydroxide
  1.4 oz. (40 g) Sodium Hydroxide
  10 oz. (284 g) Water
  12 oz. (340 g) Glycerin

Dilution Water:

  44.1 oz. (1250 g) Water
  1.8 oz. (51 g) Potassium Carbonate

Total Weight of 40% Soap Solution = 112.1oz. (3,178 g)

58% Coconut Oil Soap w/ Palm, Macadamia Nut & Castor Oils

Oils:

  20.8 oz. (590 g) Coconut Oil
  5.8 oz. (164 g) Macadamia Nut Oil
  5.4 oz. (153 g) Palm Oil
  4 oz. (113 g) Castor Oil

Lye Solution:

  6.8 oz. (193 g) Potassium Hydroxide
  1.4 oz. (40 g) Sodium Hydroxide
  10 oz. (284 g) Water
  12 oz. (340 g) Glycerin

Dilution Water:

  44.2 oz. (1,253 g) Water
  1.8 oz. (51 g) Potassium Carbonate

Total Weight of 40% Soap Solution = 112.2 oz. (3,181 g)

50% Coconut Oil Soap w/ Olive Oil

Oils:

    18 oz. (510 g) Coconut Oil

    18 oz. (510 g) Olive Oil

Lye Solution:

    6.6 oz. (187 g) Potassium Hydroxide

    1.4 oz. (40 g) Sodium Hydroxide

    10 oz. (284 g) Water

    12 oz. (340 g) Glycerin

Dilution Water:

    44 oz. (1247 g) Water

    1.8 oz. (51 g) Potassium Carbonate

Total Weight of 40% Soap Solution = 111.8 oz. (3,170 g)

CHAPTER

THICKENING LIQUID SOAP

Thickening the Soap

"Be it known that I, William Sheppard, of the city, county, and State of New York, have invented a new and Improved Liquid Soap; and I do hereby declare that the following is a full, clear, and exact description thereof, which will enable those skilled in the art to make and use the same . . . by the addition of a comparatively small quantity of common soap to a large quantity of spirits of ammonia or hartshorn is thickened to the consistency of molasses, and a liquid soap is obtained of superior detergent properties."

~ United States Patent Office, Patent No. 49,561
(DATED AUGUST 22, 1865)

One of the biggest disappointments first time liquid soapmakers express is the water-like consistency of the finished soap. Consumers have been led to equate thickness with concentration when in reality most luxuriously thick liquid cleansers are artificially thickened by the use of natural or synthetic materials. Most of these thickeners are not effective in true liquid soap due to its high pH. There are a few ways to create noticeably thicker soaps using natural ingredients.

Sodium Chloride

Sodium chloride (non-iodized table salt) works well in thickening formulas high

in soft oils. Sea salt and other types of specialty salts are not recommended due to their high content of minerals such as magnesium, calcium and copper, the same minerals that create soap scum. Instructions are easy and straightforward.

- Weigh out the soap stock to be thickened. It should be at room temperature in order to judge the thickness of the product more accurately. It is difficult to impossible to judge the thickening if the soap is hot.

- Make a 25% salt solution by dissolving 0.5 oz. (14 g) of salt in 1.5 oz. (43 g) of distilled or soft water.

- Usage rate depends upon the oils and the desired viscosity of the soap, but is approximately 0.5–1% of the salt solution to the diluted and neutralized soap stock. In this case, more is not better. There is a balance that must be met. Too little and the soap won't thicken. Too much and the soap will become cloudy and "break", becoming a thin, milky mess of ruined soap.

> Colonial soapmakers produced bar soap by salting out potassium soap. When salt (sodium chloride) is added to potassium soap an interchange of bases occurs. The chloride in salt reacts with the potassium in soap to form potassium chloride. The sodium in salt reacts with the fatty acids in soap to form sodium soap. The sodium soap rises to the surface in the form of soap curds leaving the potassium chloride, glycerin and any excess alkali in solution.

- Start at the low end of the usage rate first. Multiply the weight of the soap being diluted by 0.005 to arrive at a 0.5% usage rate of the 25% salt solution.

- Weigh out the amount of salt solution and slowly add it to the soap stock. The soap should continue to thicken over the next hour, so patience is required. If the soap has not achieved any appreciable thickening, the problem could be too much coconut oil in the formula or the soap is overly diluted and thickening with salt is not an option with this particular formula. If the soap stock has thickened, but more viscosity is desired, add another 0.5% salt solution to the soap stock and allow it to thicken for another hour. Repeat until the desired viscosity has been achieved or a total of 2% salt solution has been added. If the desired viscosity has not been reached with an overall 2% salt solution, the chances of achieving a higher viscosity with the addition of more is minimal.

Making naturally thick gel soaps without the use of thickeners is outlined in the next chapter. Other processes for thickening liquid soaps naturally by using an alcohol/lye method of soapmaking or Borax to thicken, neutralize and emulsify liquid soap can be found in *Making Natural Liquid Soaps* (Failor, 2000). These are great methods that work very well in soaps with a majority of soft oils. But how to thicken coconut oil soaps, which are the thinnest, wateriest, yet foamiest of all natural liquid soaps?

Hydroxyethylcellulose (HEC)

Hydroxyethylcellulose is naturally derived from cellulose but goes through further processing with sodium hydroxide and ethylene oxide. It is fairly natural, but to be completely honest, it is a synthetic material. It is sold under the trade names Cellosize™ by Dow Chemical and Natrasol™ by Ashland, Inc. It is the only solution I have found that works consistently to thicken all liquid soaps including 100% coconut oil soaps. Aside from thickening, it also has emulsifying, water retention and suspension properties. Liquid soap thickened with HEC will be less likely to separate or precipitate fatty acids out of solution. Its downside is its dampening effect on lather.

Usage rate is generally 0.5–1.0%. Formulations on the Dow Chemical website use percentages up to 2%. HEC comes in varying grades that range in viscosity and ease of dispersal and not all vendors carry the same grade of HEC. The usage rate to achieve the desired

Note: Dispersal of the HEC should be made at or near a neutral pH. Using the soft or distilled water intended for the further dilution of a 40% neutral soap stock to make the initial dispersal of the HEC is the most convenient.

viscosity in a formula is largely dependent upon the grade provided by the vendor. How thick is thick enough is a personal preference that must also take into consideration the bottling of the product. Getting a firm gel into a narrow opening is a laboriously slow process without proper bottling equipment. Follow the instructions below to create beautifully thick and clear shower gels.

> *Example* To dilute 100 oz. (2,835 g) of a 40% soap stock to a 35% soap stock would require 14 oz. (397 g) of additional water. The soapmaker would use the 14 oz. (397 g) of water to mix with the required amount of HEC.

- Gently warm the 40% soap stock to 140°–158°F (60°–70°C) and maintain the temperature.

- Calculate and weigh the amount of hydroxyethylcellulose required.

> *Example* Using the above example with a 1% usage rate of HEC, the soapmaker should multiply the total amount of the soap stock including the additional dilution water by 1% or 0.01 to arrive at the required amount of HEC. 100 oz. (2,835 g) + 14 oz. (397 g) = 114 oz. (3,232 g) soap stock × 1% or 0.01 = 1.14 oz. (32 g) HEC.

- Weigh the required dilution water into a pitcher or measuring cup. The water should be at room temperature or lower.

- Using the dilution water, make a slurry by adding a small amount of the water into the HEC and stirring until smooth.

Slowly whisk the HEC slurry into the water with continuous stirring.

- Once the HEC has been completely incorporated into the water, slowly add it to the warm soap stock maintaining a steady stir. A stick blender can be used but is not completely necessary. The rise in pH and temperature will increase the hydration of the HEC.

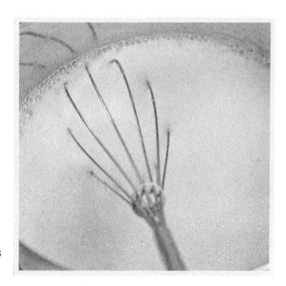

- Continue stirring a few minutes after the soap has become thick and homogenous (no appearances of globules or beads). Total time should take 5–25 minutes.

- Place a cover over the pot and allow any foam to subside. Once the foam subsides, you should have a beautifully thick and luxurious shower gel ready for fragrance and/or color.

CHAPTER

HOW TO MAKE SOAP GELS AND JELLIES

Soap Gels and Jellies

"By soft soap is understood a soap prepared chiefly from potash lye and oil, of a soft, salve-like consistency, and containing much more water chemically fixed than hard soap. It is not a simple alkaline sebate, but a solution of an alkaline sebate in a mixture of carbonated and caustic lyes. As neutral potassium oleate yields a viscid, gummy, and turbid product, it is necessary to add to it, in order to obtain the transparent soap of commerce, a solution of potassium hydroxide and carbonate, or to substitute for the latter one of allied salts. By depositing itself between the atoms of soap, the potassium carbonate, added in suitable proportion, breaks up the viscidity of the soap and forms it into a plastic, transparent mass, while it also possesses the property of combining water with the soap."

~ THE SOAP MAKER'S HANDBOOK OF MATERIALS, PROCESSES AND RECEIPTS FOR EVERY DESCRIPTION OF SOAP . . . (DEITE, ENGELHARDT, WILTNER, 1912)

Soap Gels and Jellies are made using a combination of the paste and no paste soapmaking method. The main difference in the process is the majority oil in the formulas will be unsaturated oils such as olive, high oleic sunflower, high oleic safflower etc. This is where the soapmaker will need to understand the effects the oils have on the viscosity of the finished soap.

The extreme solubility of coconut and castor reduce their role in gel soapmaking to a supporting actor. Too much coconut oil or in combination with castor oil will create a liquid soap rather than a gel soap. Higher percentages of coconut oil may be manipulated into the formula by using a percentage of sodium hydroxide in conjunction with the potassium hydroxide.

> Experience with making paste soaps before tackling Gel and Jelly Soap is strongly recommended.

Gelled soap requires both the potassium carbonate and glycerin in order to make a smooth transparent gel of acceptable consistency. Without these two ingredients, the soap maintains its Vaseline paste consistency until it reaches a saturation point at which time it begins to separate into layers. Glycerin without potassium carbonate creates a transparent weepy paste. Potassium carbonate without glycerin creates a very plastic gel that is difficult to manage and requires more dilution.

An important consideration when making soap gels and jellies is the quality of the stick blender. The stick blender should be powerful enough to create a vortex within the solution when blending the soap. Excessive foaming, created by the necessity of moving the stick blender all around the pot to incorporate the ingredients, will break the soap into layers. Achieving gel soap at this point is almost impossible unless the stick blender is exchanged for a more efficient one and the separation can be blended back together until it no longer separates.

Gel Soapmaking Instructions:

Step 1 **Safety First!** Don goggles, gloves protective mask and clothing!

Step 2 **Gather Thoughts and Materials:** Make sure you have everything you need.

Step 3 **Weigh and Melt the Oils:** Ideal temperature should be between 180°F (82°C)–190°F (88°C) depending on the oils. If a higher percent of coconut oil is in the blend, maintain a temperature of 180°F (82°C).

Step 4 **Prepare the Dilution Water and the Potassium Carbonate:** Weigh the potassium carbonate, the dilution water and any other additives called for in the recipe and heat to boiling. Remove from heat and set aside. Approximately half of the dilution water will be added at trace and the other half will be added a little at a time over the course of the first hour. Soaps high in soft oils are more prone to separation. Adding the dilution water gradually helps maintain an emulsion.

Step 5 **Prepare the Glycerin:** Weigh the glycerin and heat in the microwave or on the stove until hot but not boiling. Cover and keep warm.

Step 6 **Prepare the Lye Solution:** All of the recipes in this section include potassium hydroxide. Sodium hydroxide is an optional ingredient that can be calculated using the instructions in "Chapter 15: Formulating Liquid Soap Recipes."

In 100% soft oil recipes the use of sodium hydroxide will create jelly soap with the texture and consistency of Jell-O®.

The water portion of the lye solution is discounted to allow the addition of glycerin into the lye solution. The glycerin will speed the saponification of the oils. Using it in the lye solution is much more effective than adding it later in the process.

- Measure the water amount of the lye solution and set aside.

- Measure the amount of hydroxides and add to the water, stirring until dissolved.

- Add the hot glycerin into the dissolved lye solution. Make sure the lye is completely dissolved before adding the hot glycerin! Failure to do so may result in a volcano which could cause serious burns as well as damage to surrounding surfaces!

Step 7 **Stir the Lye Solution into the Oils:** Once the lye solution has been completely dissolved and the glycerin added to the mixture, it should be at a good temperature to add to the oils. Slowly pour the lye solution into the oils and check the temperature. If it is below 180°F (82°C), warm it up until the desired temperature is reached. If it is above 190°F (88°C), allow the oil/lye solution to cool down before stick blending to avoid boil over. Once the desired range of temperature is achieved introduce the stick blender and blend until trace.

Step 8 **Bring to Light Trace:** Avoiding as much foaming as possible by using higher speeds to create a vortex, bring the soap to light trace. The definition of trace using the gel method diverges from the paste method and more closely resembles thick trace in cold process soap, with an exception! Due to the glycerin in the lye solution, the oil/lye mixture will be translucent instead of opaque and creamy. Trace will look more like a gel than a pudding (compare thickening sauces with cornstarch rather than flour). A good way to gauge trace is when the impression of the stick blender clearly remains in the soap solution when it is removed.

Step 9 **Add the Dilution Water and/or Additives:** Sometimes separation may occur at trace and the emulsion will have a grainy, applesauce texture. When this happens, add a small amount of the dilution water while stick blending to bring it back to an emulsion. Continue adding water if the soap solution becomes noticeably thicker regardless of whether or not the separation has been resolved. Once the soap solution has reached trace and is creamy and uniform, slowly add half of the dilution water called for in the recipe.

Step 10 **Cook the Soap:** The trick to making gel soap is maintaining a gelled state. It is important to avoid a stiff, unstirrable paste that will have to be broken up to be diluted. Add the remainder of the dilution water within the first hour of cooking. Add small (1—2 oz. (28—57 g)) increments of distilled or soft water throughout the process to replace any water lost through evaporation and to maintain a fluid paste.

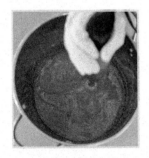

Cooking on top of the stove is not recommended. Crock pots have a tendency to get too warm and may cause the soap to separate. Allowing the soap to remain in a separated state can affect the gel. If the soap does separate stick blend until it comes back into an emulsion. Cooking in the oven or a double boiler appears to give the best results. Maintain a temperature between 185°–195°F (85°–91°C) and periodically stir the soap as it cooks. As it cooks it will stiffen into a firm gel or soft paste. Avoid a thickening of the upper part of the paste by keeping the soap solution tightly covered while cooking.

Continue cooking until the soap becomes transparent. Temperature and ingredients will determine the amount of time required, but soft oil soaps take much longer to cook than coconut oil soaps.

Gel soap in its concentrated form suspends the bubbles and traps them within the soap. Complete transparency will be difficult to gage. Because a concentrated soap cooks more efficiently, it is important to reach the transparent stage at a higher concentration of soap. Gel recipes are calculated with a 40% soap concentration during the cook.

Step 11 **Check for Neutrality:** Before diluting further, check for cloudiness in the soap by dissolving a small sample in warm distilled or soft water. If it remains clear, the soap is ready for further dilution. If the soap is cloudy, test for free alkali with phenolphthalein. If the solution turns pink, continue the cook and test with phenolphthalein again until a clear solution is achieved. If the soap remains colorless, follow the directions for neutralizing excess oils on page 42. Continue the cook and test again until a clear solution is achieved.

Step 12 **Dilute the Soap Gel:** While the gel is still warm, dilute to a 35% solution using the chart below. Add the water slowly and stir between additions. A stick blender may be used if desired, but is not necessary and will take longer for the foam to disperse. Allow the soap to cool in order to determine the viscosity of the soap. At this point, soft or distilled water can easily be added to the room temperature soap to achieve the desired thickness. Excessive stringiness (or snottiness, you'll know what I mean when you see it) in the soap is a sign of too high a soap concentration. The soap can be used in this higher concentration without concern of separation, but it is difficult to work with and has an undesirable feel. Continue diluting until a portion of soap pinched between your thumb and forefinger no longer strings.

Gel Soap Dilution Chart

Desired Concentration of Soap	Starting @ 40% Soap Multiply by	Starting @ 35% Soap Multiply by	Starting @ 30% Soap Multiply by
35%	0.14	—	—
30%	0.33	0.17	—
25%	0.60	0.40	0.2
20%	1.0	0.75	0.50
15%	1.67	1.33	1.00

Step 13 **Allow the Soap to Settle:** The soap will continue to neutralize on its own over the next several days. Any lumps formed during the cook will also dissolve within the gel.

Step 14 **Neutralize the Soap:** Warm the soap gel and neutralize with citric acid according to the directions on page 34. Neutralizing with stearic acid will not work with this method.

Step 15 **At this point the soap can be used as is or** colored (Chapter 10) and scented (Chapter 11) as desired. Unlike liquid soaps, gelled soaps are less likely to precipitate excess fatty acids or unsaponifiables out of solution, therefore sequestering the soap longer is not required unless highly diluted.

Step 16 **Bottling:** Unless the soapmaker has a piston filling machine for filling the soap into bottles, it may be necessary to further dilute the gel or slightly warm the soap in order to pour into the containers, however if the soap is too hot, it will warp the bottles. Using a pastry bag to fill the bottles is another option. Pet bottles should have narrow sides if using a dispensing cap with stiff gels because of their rigid construction.

CHAPTER 9

SOAP GELS AND JELLIES RECIPES

Gel & Jelly Soap Recipes

"Oil soaps, sometimes designated as jelly soaps, are generally sold in bulk to factories, institutions, office buildings, etc. They are used for general cleaning of floors, walls and woodwork. Noteworthy features of oil soaps are that they dissolve easily, are unguent due to bland oils and glycerol content, are harmless to most surfaces, rinse quickly and leave no residue"

~ Sanitary Products (SCHWARTZ, 1943)

85% OLIVE OIL SOAP W/ COCONUT OIL

Oils:

30.6 oz. (868 g) Olive Oil

5.4 oz. (153 g) Coconut Oil

Lye Solution:

7.8 oz. (221 g) Potassium Hydroxide

9.9 oz. (281 g) Water

12 oz. (340 g) Glycerin

Dilution Water:

43.8 oz. (1,242 g) Water (add half at trace and the other half during the cook)

1.8 oz. (51 g) Potassium Carbonate

Total Weight of 40% Soap Solution = 111.3 oz. (3,155 g)

80% Olive Oil Soap w/ Coconut and Castor Oils

Oils:

28.8 oz. (817 g) Olive Oil

5.4 oz. (153 g) Coconut Oil

1.8 oz. (51 g) Castor Oil

Lye Solution:

7.8 oz. (221 g) Potassium Hydroxide

9.9 oz. (281 g) Water

12 oz. (340 g) Glycerin

Dilution Water:

43.8 oz. (1,242 g) Water (add half at trace and the other half during the cook)

1.8 oz. (51 g) Potassium Carbonate

Total Weight of 40% Soap Solution = 111.3 oz. (3,155 g)

75% High Oleic Sunflower Oil Soap w/ Coconut Oil, Cocoa Butter And Castor Oil

Oils:

27 oz. (765 g) High Oleic Sunflower Oil

1.8 oz. (51 g) Cocoa Butter

3.6 oz. (102 g) Coconut Oil

3.6 oz. (102 g) Castor Oil

Lye Solution:

7.8 oz. (221 g) Potassium Hydroxide

9.9 oz. (281 g) Water

12 oz. (340 g) Glycerin

Dilution Water:

43.8 oz. (1,242 g) Water (add half at trace and the other half during the cook)

1.8 oz. (51 g) Potassium Carbonate

Total Weight of 40% Soap Solution = 111.3 oz. (3,155 g)

90% Olive Oil Soap w/ Coconut Oil

This is my very favorite gel soap recipe. It makes a nice thick gel with good lathering properties in hard and soft water! It is also one of the easier gels to make.

Oils:

32.4 oz. (919 g) Olive Oil

3.6 oz. (102 g) Coconut Oil

Lye Solution:

7.8 oz. (221 g) Potassium Hydroxide

9.9 oz. (281 g) Water

12 oz. (340 g) Glycerin

Dilution Water:

43.8 oz. (1,242 g) Water (add half at trace and the other half during the cook)

1.8 oz. (51 g) Potassium Carbonate

Total Weight of 40% Soap Solution = 111.3 oz. (3,155 g)

75% Olive Oil Soap w/ Avocado and Castor Oils

Oils:

27 oz. (765 g) Olive Oil

5.4 oz. (153 g) Avocado Oil

3.6 oz. (102 g) Castor Oil

Lye Solution:

7.4 oz. (210 g) Potassium Hydroxide

9.7 oz. (275 g) Water

12 oz. (340 g) Glycerin

Dilution Water:

43.4 oz. (1,230 g) Water (add half at trace and the other half during the cook)

1.8 oz. (51 g) Potassium Carbonate

Total Weight of 40% Soap Solution = 110.3 oz. (3,127 g)

85% High Oleic Sunflower Oil Soap w/ Sweet Almond Oil

Oils:

30.6 oz. (868 g) High Oleic Sunflower Oil

5.4 oz. (153 g) Sweet Almond Oil

Lye Solution:

7.6 oz. (216 g) Potassium Hydroxide

9.8 oz. (278 g) Water

12 oz. (340 g) Glycerin

Dilution Water:

43.6 oz. (1,236 g) Water (add half at trace and the other half during the cook)

1.8 oz. (51 g) Potassium Carbonate

Total Weight of 40% Soap Solution = 110.8 oz. (3,141 g)

90% OLIVE OIL SOAP W/ CASTOR OIL

An extremely mild and moisturizing soap.

Oils:

32.4 oz. (919 g) Olive Oil

3.6 oz. (102 g) Castor Oil

Lye Solution:

7.4 oz. (210 g) Potassium Hydroxide

9.7 oz. (275 g) Water

12 oz. (340 g) Glycerin

Dilution Water:

43.4 oz. (1,230 g) Water (add half at trace and the other half during the cook)

1.8 oz. (51 g) Potassium Carbonate

Total Weight of 40% Soap Solution = 110.3 oz. (3,127 g)

75% Olive Oil Soap w/ Palm, Coconut & Castor Oils

Oils:

27 oz. (765 g) Olive Oil

1.8 oz. (51 g) Palm Oil

3.6 oz. (105 g) Coconut Oil

3.6 oz. (105 g) Castor Oil

Lye Solution:

7.7 oz. (218 g) Potassium Hydroxide

9.8 oz. (278 g) Water

12 oz. (340 g) Glycerin

Dilution Water:

43.5 oz. (1,233 g) Water (add half at trace and the other half during the cook)

1.8 oz. (51 g) Potassium Carbonate

Total Weight of 40% Soap Solution = 110.6 oz. (3,136 g)

100% Olive Oil

A very stiff gel can be achieved, great for dispensing in tottle bottles as a mild face wash.

Oils:

36 oz. (1021 g) Olive Oil

Lye Solution:

7.5oz. (213 g) Potassium Hydroxide

9.8 oz. (278 g) Water

12 oz. (340 g) Glycerin

Dilution Water:

43.5 oz. (1,233 g) Water (add half at trace and the other half during the cook)

1.8 oz. (51 g) Potassium Carbonate

Total Weight of 40% Soap Solution = 110.6 oz. (3,136 g)

65% OLIVE OIL SOAP w/ CASTOR & COCONUT OILS

The next four recipes take the gel to its outermost limits with their high percentage of coconut oil. The castor oil in this recipe will thin the gel even more. Using a lower dilution rate or a combination of potassium and sodium hydroxide will increase the viscosity.

Oils:

23.4 oz. (664 g) Olive Oil

9 oz. (255 g) Coconut Oil

3.6 oz. (102 g) Castor Oil

Lye Solution:

8.1 oz. (230 g) Potassium Hydroxide

10 oz. (284 g) Water

12 oz. (340 g) Glycerin

Dilution Water:

44.1 oz. (1,250 g) Water (add half at trace and the other half during the cook)

1.8 oz. (51 g) Potassium Carbonate

Total Weight of 40% Soap Solution = 112 oz. (3,175 g)

70% Olive Oil Soap w/ Coconut Oil

This recipe makes a mild soap with a luxurious lather that more than compensates for its lower viscosity.

Oils:

   25.2 oz. (714 g) Olive Oil
   10.8 oz. (306 g) Coconut Oil

Lye Solution:

   8.2 oz. (233 g) Potassium Hydroxide
   10.1 oz. (286 g) Water
   12 oz. (340 g) Glycerin

Dilution Water:

   44.2 oz. (1,253 g) Water (add half at trace and the other half during the cook)
   1.8 oz. (51 g) Potassium Carbonate

Total Weight of 40% Soap Solution = 112.3 oz. (3,184 g)

75% Olive Oil Soap w/ Coconut Oil

Oils:

   27 oz. (765 g) Olive Oil
   9 oz. (255 g) Coconut Oil

Lye Solution:

   8.2 oz. (232 g) Potassium Hydroxide
   10.1 oz. (286 g) Water
   12 oz. (340 g) Glycerin

Dilution Water:

    44.2 oz. (1,253 g) Water (add half at trace and the other half during the cook)

    1.8 oz. (51 g) Potassium Carbonate

Total Weight of 40% Soap Solution = 112.3 oz. (3,184 g)

60% OLIVE OIL SOAP W/ COCONUT, CASTOR & PALM OILS

An example of using both potassium and sodium hydroxide in a formula higher in coconut and castor oils to achieve a thicker soap. The palm oil also helps to increase the viscosity and stabilize the lather.

Oils:

    21.6 oz. (612 g) Olive Oil

    9 oz. (255 g) Coconut Oil

    3.6 oz. (102 g) Castor Oil

    1.8 oz. (51 g) Palm Oil

Lye Solution:

    6.1 oz. (173 g) Potassium Hydroxide

    1.3 oz. (37 g) Sodium Hydroxide

    9.7 oz. (275 g) Water

    12 oz. (340 g) Glycerin

Dilution Water:

    43.4 oz. (1,230 g) Water (add half at trace and the other half during the cook)

    1.8 oz. (51 g) Potassium Carbonate

Total Weight of 40% Soap Solution = 110.3 oz. (3,126 g)

CHAPTER

COLORING LIQUID SOAPS

"Besides indigo or ultramarine blue, chlorophyll, as well as the very intense hot water-soluble greens F and M are also used for coloring soft soaps green."

~ AMERICAN SOAP MAKER'S GUIDE
(MERBOTT, STANISLAUS, 1928)

Liquid soaps are naturally beautiful. Clear and sparkling colors from pale champagne, olive green, gold or amber are created by the oils in the formula. Essential oils such as lemongrass and folded citrus oils contribute color hues of their own. Achieving other than the natural palette of colors provided by the ingredients requires the use of water soluble dyes or oil infused herbs. Oil soluble pigments and micas will precipitate out of the liquid soap so are not recommended.

D&C and FD&C Dyes

Water based dyes include D&C and FD&C dyes and are the best option for coloring liquid soaps. D&C stands for drug and cosmetics and FD&C stands for food, drug and cosmetics making it easy to understand which dyes are approved for a particular application. Water based dyes can be tricky and the color may mutate in alkaline solutions so it is important to test the colorant first before applying it to a whole batch of soap. The natural golden color of the soap and the addition of essential oils that darken also make it very difficult to achieve particular colors. Many of the soap vendors will have already tested the colorants and will have instructions on how best to use them in cold process soap. Those same rules will help in selecting colors that will give dependable results in liquid soaps.

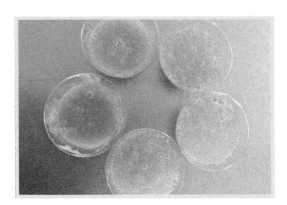

The easiest way to play with colors when first starting out is with Wilton™ Icing Colors or McCormick™ Food Coloring. The basic McCormick set found in every grocery store has too much yellow in the reds and blues so that most of the soap is either orange or green, but both brands can be found in major craft stores in a wide array of colors to suit any palette. If using the Wilton™ Icing colors, thin them down first with distilled or soft water before adding to the soap. This will prevent incorporating any small dried flakes of color that sometimes collect around the top of the pot and refuse to dissolve in the soap. Begin by using minute amounts of color because a little goes a long way.

Natural Colorants

Coloring soap with infused herbs is tricky. Any natural colorant that can be infused in oil such as alkanet root (purple or muddy grey), turmeric (deep yellow), burdock leaf (green) annatto seed(bright yellow), indigo (blue), paprika (peach), red sandalwood powder (maroon) can be infused with the oil, strained and added as a part of the soap stock.

The problem lies in the undependable color variation from batch to batch. The color achieved one time may be entirely different the next even though the soapmaker scrupulously measured both recipes.

Another issue the soapmaker must consider is if the soap is legally considered a cosmetic. Only colorant additives approved by the FDA for use in cosmetics may be used to color the soap.

If, however the soapmaker wishes to attempt coloring the soap naturally, the same procedure as outlined in "Chapter 12: Incorporating Additives," will work for steeping colored infusions. Limit the amount of the herb to ¼–2 teaspoons rather than adding by the cupful. Precise measurements depend upon the strength of the herb, the amount of soap to be colored and the depth of color desired.

Pearl Soap

Pearl soap is a translucent to opaque soap with an iridescent silver sheen. Making pearl soap can be as easy as adding a pearling agent to the product. While the pearlizer doesn't add any color, it does add dramatic depth and luster to liquid soap and can conveniently be used to camouflage cloudiness caused by waxes, butters and additives. Colors can also be added with the pearlizer to create even more interest to the soap. There are two products available to the handcrafted soapmaker.

1. EZ Pearl is 100% Glycol Distearate (a diester of ethylene glycol). It comes as a waxy flake and must be incorporated into the product at a temperature above 149°F (65°C) and allowed to cool slowly. Because it increases the viscosity of the product, testing first in smaller samples is highly recommended. Follow the manufacturer's recommended usage rate.
2. Pearling agents formulated with Glycol Distearate in combination with various surfactants are designed for ease of dispersal in cold solutions. Added benefits include the lather and detergency booster of the surfactants especially useful in soap shampoos and bubble baths. The downside, the synthetic surfactant in the pearlizer will also be in the soap as well as on the label. Usage rate depends on the concentration as well as the supplier. Viscosity may increase in some formulas.

If the soapmaker is using potassium carbonate in the soap formula, the soap will be more of a translucent pearl than opaque and may require a higher usage rate of the product.

CHAPTER

FRAGRANCING SOAPS NATURALLY

Blending Essential Oils for Soap

"Successful soap perfuming therefore requires a knowledge of the durability of each individual raw material when in contact with the soap and any of its constituent impurities."

~ *Perfumes, Cosmetics and Soaps*, Vol. II, (Poucher, 1932)

Once upon a time, if an essential oil was for sale at an affordable price, it was fair game for blending and using in any and all of the products placed upon the market without restrictions. It was up to the perfumer to proceed cautiously regarding oils that were known allergens or harmful to the health of the individual or the environment. Things have changed radically over the last decade and some of the greatest changes are the rules and regulations set forth by the EU Cosmetic Directive, Annex III, 7th Amendment and the International Fragrance Association (IFRA).

The EU Cosmetic Directive, Annex III, 7th Amendment requires for cosmetics sold in the European Union (EU), the presence of any of 26 fragrance components identified as an important cause of allergy be listed in the ingredients to ensure customers are adequately informed.

The IFRA standards have set forth the recommended safe use of fragrance ingredients based on research by the Research Institute for Fragrance Material (RIFM). It is a self-regulating membership based system of the industry. The IFRA standards have the force of law in the European Union whether the company or individual manufacturing fragrances is a member or not. So far it continues to be voluntary within the United States. Whether a member or not, and however voluntary, the IFRA standards are held to be the most accepted and recognized standards of the industry on the safe use of fragrance ingredients and should be considered good manufacturing practices for all fragrance manufacturers, including the handcrafted soapmaker. This will most likely be the argument for the plaintiff in any lawsuit against someone blatantly ignoring IFRA restrictions in their products.

An RIFM panel of experts (REXPAN) comprised of independent dermatologists, toxicologists, pathologists and environmental scientists evaluate data collected by IFRA on usage, concentration, exposure levels and importance of a chemical to perfumery and make a determination of safe usage level based on any risk factors. Every two years new amendments are made to the standards and posted on the IFRA website (www.ifraorg.org). As of this writing, The IFRA has listed 174 substances which have been banned or have restricted their uses in fragrance products and these restrictions depend in large part on the application. Specifications have also been placed on oils that may have ingredients that have a tendency to oxidize. It has been determined that oxidation is more likely to cause allergic reactions. Oils containing pinacea derivatives which include fir, balsam and pine essential oils, oils rich in linalool such as rosewood, coriander and ho wood and oils rich in limonene such as citrus oils have limitations according to the peroxide levels in the product. These oils should be preserved against oxidation by the addition of an antioxidant at the time of production. The soapmaker should store them appropriately and strictly adhere to the use by date as suggested by the manufacturer.

The IFRA has divided the application of fragrances into 11 categories with guidelines for each. Use of essential oils containing any of these components is restricted accordingly. To address all of the essential oils affected and the usage level by category for each product is a book unto itself. The information provided below is based on the 47th Amendment of the IFRA standards for Categories 9 and 10 for the restrictions and specifications affecting some of the most common essential oils used by the handcrafted soap industry. It is a compilation of information found on the IFRA and other third party websites and is intended for general information only. The soapmaker should not rely on any information gathered here as authoritative or a substitute for IFRA Certification papers which are available to the consumer through the manufacturer or distributor of the essential oil. The IFRA standards change frequently as research and study of the various essential oils evolve. It is important for the soapmaker to keep abreast of any of those changes.

- Category 9 includes rinse off products such as soap, body washes, bath salts, bath oils and hair conditioners. **Maximum concentration of any fragrance in the product may not exceed 5% in the finished product.**

- Category 10 includes household cleaners such as laundry detergents, fabric softeners and hard surface cleaners including concentrated cleaners. **Maximum concentration of any fragrance in the product may not exceed 2.5% in the finished product.**

Any other products made by the handcrafted soapmaker that do not fit into these two categories will require further investigation by the soapmaker. A quick reference chart of 25 essential oils along with usage guidelines can be found in Appendix E.

> The math in determining the usage rate of essential oils with restrictions can be mind boggling to some. The best tool I found was a fragrance allergen calculator found online. There are several out there with clear instructions on use. I would encourage those who would rather avoid the math to give them a try.

Twenty-Five Essential Oils for Blending in Soap

A summary of twenty five essential oils commonly used by the handcrafted soapmaker along with any restrictions placed upon them and a brief description of their benefits.

Bergamot (*Citrus bergamia*)

Bergamot oil is derived from the fruits of a small tree. The fruits ripen from green to yellow and resemble oranges, although smaller. Bergamot has a flowery yet citrus undertone and is excellent as a modifier and sweetener in blending with citrus oils and colognes. It has a very effective top note and an ability to anchor more fleeting oils. Use to enhance and strengthen, lift and refresh light and airy blends. Follow IFRA guidelines for use in other than wash-off products.

IFRA Restrictions: No restrictions in wash-off products. Bergamot is a furocoumarin containing essential oil. Restrictions apply for leave on applications unless it has been rectified bergapten free.

IFRA Specifications: Bergamot oil contains the following chemical components specified under IFRA guidelines: d-Limonene, limit peroxide levels to 20 mmol/l.

EU Cosmetic Directive, Annex III, 7th Amendment: Bergamot oil contains the following chemical components that must be listed on the label if the percentage of the chemical component in the final product is in excess of 0.001% in leave on products and 0.01% in wash off products. Bergamot Oil Distilled: Citral < 0.4%; Bergamot Oil Expressed: Citral < 0.7%.

Bergamot oil may be used up to 100% of the fragrance concentrate at a usage rate of up to 5% of the total weight in the finished product under category 9.

Cassia (*Cinnamomum cassia*)

Although the restrictions placed on cassia by the IFRA standards make it almost not worthy of mentioning, it's the only oil available to the soapmaker at a reasonable price that gives the spicy cinnamon flavor to a blend.

IFRA Restrictions: Cassia oil contains the following chemical components restricted under IFRA guidelines: Eugenol < 0.5%, Methyl eugenol < 0.1%, Coumarin < 4%, Benzaldehyde < 2%, Benzyl benzoate < 1%, Cinnamic alcohol < 1%, Cinnamic aldehyde < 90%.

IFRA Specifications: None at this time.

EU Cosmetic Directive, Annex III, 7th Amendment: Cassia oil contains the following chemical components that must be listed on the label if the percentage of the chemical component in the final product is in excess of 0.001% in leave on products and 0.01% in wash off products: Eugenol, Benzyl alcohol, Benzyl benzoate, Cinnamyl alcohol, Cinnamic aldehyde, Coumarin, Limonene and Linalol.

Due to its high cinnamic aldehyde content, cassia oil is restricted to a maximum of 0.05% in the finished product under Category 9 and 10 of the IFRA Guidelines. This would equate to a fragrance blend concentrate containing up to 5% cassia oil with a usage rate of up to 1% of the fragrance concentrate in the total weight of the finished product. With a usage rate of up to 5% of a fragrance concentrate in the finished product, the amount of cassia oil would be reduced to a maximum of 1% in a fragrance concentrate.

Consideration must be given to the following IFRA restricted ingredients in cassia oil if used in combination with other essential oils or fragrance components also containing these ingredients. The combined percentage of the restricted ingredient in all fragrance components within the formula must be calculated. The IFRA restricted ingredients are listed showing the maximum percentage allowed in the finished product.

Chemical Isolate	Maximum Percentage Allowed in Finished Product for IFRA Category 9 & 10
Benzaldehyde	3%
Cinnamic Alcohol	0.4%
Cinnamic Aldehyde	0.05%
Eugenol	0.5%
Methyl Eugenol	0.001%

Atlas Cedarwood (*Cedrus atlantica*)

Atlas or Atlantic cedarwood is thought to have originated from the Lebanon cedar. Lebanon cedar was used extensively by the ancient Egyptians, Greeks and Phoenicians as well as many others in religious ceremonies, cosmetics, perfumery and embalming for the aroma's alleged ability to preserve, protect and defend.

Atlas cedarwood is a true cedar unlike the Texas and Virginia cedar which is from the juniper species. It is one of the cedarwood oils specified as a fixative in perfumery. It is used extensively in soap perfumery as a substitute for sandalwood. Atlantic cedarwood has an uncanny ability to meld its odor with other oils when blended, thereby making it a good choice in blending a variety of fragrances.

IFRA Restrictions: None at this time.

IFRA Specifications: None at this time.

EU Cosmetic Directive, Annex III, 7th Amendment: None at this time.

Atlas cedarwood oil may be used up to 100% of the fragrance concentrate at a usage rate of up to 5% of the total weight in the finished product under category 9.

Clary Sage (*Salvia sclarea*)

Clary sage plays a special role in perfumery. It is an excellent blender, modifier and fixator and is suitable for any scent blend. As a fixator, its high flash point and specific gravity help raise the flashpoint of the other essential oils in the blend. As a modifier, it aids in smoothing the harsh edges of other essential oils. The results are not immediate but require standing and maturing for at least a week to realize the more pleasant and tenacious representation clary sage brings to the blend.

IFRA Restrictions: Clary sage oil contains the following chemical components restricted under IFRA guidelines: Geraniol < 2.2%, trans-2-Hexenal traces < 0.1%, Coumarin 0.1%.

IFRA Specifications: Clary sage oil contains the following chemical components specified under IFRA guidelines: Linalol, limit peroxide levels to 20 mmol/l.

EU Cosmetic Directive, Annex III, 7th Amendment: Clary sage oil contains the following chemical components that must be listed on the label if the percentage of the chemical component in the final product is in excess of 0.001% in leave on products and 0.01% in wash off products. Limonene < 3%, Linalol 35%, Geraniol < 2.2%.

Due to traces of the chemical component trans-2-Hexenal, clary sage oil is restricted to a maximum of 2% in the finished product under Category 9 and 10 of the IFRA Guidelines. This would equate to a fragrance blend concentrate containing up to 100% clary sage oil with a usage rate of up to 2% of the fragrance concentrate in the total weight of the finished product. With a usage rate of 5% of the fragrance concentrate in the finished product, the amount of clary sage oil would be reduced to a maximum of 40% in the fragrance concentrate. If other essential oils or fragrance components within the fragrance concentrate also contain trans-2-Hexanol, the combined percentage of all trans-2-Hexenal containing ingredients must not exceed 0.002% in the finished product.

Clove Bud (*Szygium aromaticum*)

Spicy and warm, clove bud is used in soap perfumery as a modifier and sweetener for giving a pleasing sweetness and intensity to a blend. It is considered by some to be a fixative due to its tenacity and high flashpoint value. Clove bud oil is the least sensitizing of the clove family and the better choice for blending.

IFRA Restrictions: Clove bud oil contains the following chemical components restricted under IFRA guidelines: Eugenol < 90%, Isoeugenol < 0.1%, Methyl Eugenol Traces < 0.1%.

IFRA Specifications: None at this time.

EU Cosmetic Directive, Annex III, 7th Amendment: Clove bud oil contains the following chemical components that must be listed on the label if the percentage of the chemical component in the final product is in excess of 0.001% in leave on products and 0.01% in wash off products: Eugenol < 90%.

Due to its high eugenol content, clove bud oil is restricted to a maximum of 0.55% in the finished product under Category 9 and 10 of the IFRA Guidelines. This would equate to a fragrance blend concentrate containing up to 55% clove bud oil with a usage rate of up to 1% of the fragrance concentrate in the total weight of the finished product. With a usage rate of up to 5% of the fragrance concentrate in the finished product, the amount of clove bud oil would be reduced to a maximum of 11% in the fragrance concentrate.

Consideration must be given to the following IFRA restricted ingredients in clove bud oil if used in combination with other essential oils or fragrance components also containing these ingredients. The combined percentage of the restricted ingredient in all fragrance components within the formula must be calculated. The IFRA restricted ingredients are listed showing the maximum percentage allowed in the finished product.

Chemical Isolate	Maximum Percentage Allowed in Finished Product for IFRA Category 9 & 10
Eugenol	0.5%
Isoeugenol	0.02%
Methyl Eugenol	0.001%

Eucalyptus (*Eucalyptus globulus*)

Historically, infusions of eucalyptus leaves were used by the Aboriginals to treat sore muscles, fever and congestion. It has the distinction of being the first industry in Australia, and is a traditional household remedy for treating aching joints, sprains, burns, wounds, insect bites and headaches.

Eucalyptus oil is used extensively in soap, detergent and toiletry preparations for its refreshing, light scent. It is an excellent solvent and is widely used in spot and stain removers. Due to its effectiveness in killing bacteria, it is also used in disinfectants and household cleansers.

IFRA Restrictions: None at this time.

IFRA Specifications: None at this time.

EU Cosmetic Directive, Annex III, 7th Amendment: Eucalyptus oil contains the following chemical components that must be listed on the label if the percentage of the chemical component in the final product is in excess of 0.001% in leave on products and 0.01% in wash off products. D Limonene < 15%, Geraniol < 0.5%, Linalol < 0.5%.

Eucalyptus oil may be used up to 100% of the fragrance concentrate at a usage rate of up to 5% of the total weight in the finished product under category 9.

Lemon Eucalyptus (*Eucalyptus citriodora*)

Lemon Eucalyptus has a fresh lemon scent very similar to citronella and is a good replacement due to the lack of methyl eugenol in its chemical composition. Like citronella, it is used extensively as an insect repellant. It has antifungal and antibacterial properties and is used in skin care to treat fungal infections, dandruff and herpes. In perfumery, it is used as a modifier, blender and sweetener and is especially useful in sweetening lemongrass and lightening the heaviness of palmarosa.

IFRA Restrictions: Lemon eucalyptus contains the following chemical components restricted under IFRA guidelines: Geraniol < 0.5%, Citronellol < 8%.

IFRA Specifications: None at this time.

EU Cosmetic Directive, Annex III, 7th Amendment: Lemon Eucalyptus oil contains the following chemical components that must be listed on the label if the percentage of the chemical component in the final product is in excess of 0.001% in leave on products and 0.01% in wash off products. Citronellol < 8%, Geraniol < 0.5%, D Limonene < 0.5%, Geraniol < 0.5%, Linalol < 1.0%.

Lemon Eucalyptus oil may be used up to 100% of the fragrance concentrate at a usage rate of up to 5% of the total weight in the finished product under category 9.

Geranium (*Pelargonium graveolens*)

Referred to as the happiness plant, geranium is indispensable in all skin care products for its ability to treat oily as well as dry skin. Its regenerative properties make it especially useful for aged, mature skin. Geranium is excellent in the perfumery of soap adding floral sweet tones to almost any blend. "No well-perfumed soap of good quality is complete without a liberal quota of this raw material." *Perfumes, Cosmetics and Soaps* (Poucher, 1932). Sweetening the geranium with mint will give a more rose-like odor with less powdery undertone.

IFRA Restrictions: Geranium oil contains the following chemical components restricted under IFRA guidelines: Citral < 1.50%, Citronellol < 28%, Geraniol < 20%.

IFRA Specifications: Geranium oil contains the following chemical components specified under IFRA guidelines: Limonene 0.17%, Linalool < 5%, limit peroxide levels to 20 mmol/l.

EU Cosmetic Directive, Annex III, 7th Amendment: Geranium oil contains the following chemical components that must be listed on the label if the percentage of the chemical component in the final product is in excess of 0.001% in leave on products and 0.01% in wash off products. Citral < 7%, Citronellol < 28%, Geraniol < 20%, Limonene < 0.17%, Linalol < 5%.

Geranium oil may be used up to 100% of the fragrance concentrate at a usage rate of up to 5% of the total weight in the finished product under category 9.

Egyptian (Rose) Geranium (The Lebermuth Co., Inc.)

Egyptian (Rose) Geranium is a manufactured fragrance oil blended from natural isolates and essential oils. It serves as an excellent alternative to true geranium essential oil in soap formulas due to its attractive price and natural, plant based ingredients.

IFRA Restrictions: Egyptian (Rose) Geranium contains the following chemical components restricted under IFRA guidelines: Citral 0.08%, Geraniol 0.36%, Citronellol 0.83%.

IFRA Specifications: Egyptian (Rose) Geranium oil contains the following chemical components specified under IFRA guidelines: Limonene 0.01%, Linalool 0.16%, limit peroxide levels to 20 mmol/l.

EU Cosmetic Directive, Annex III, 7th Amendment: Egyptian (Rose) Geranium contains the following chemical components that must be listed on the label if the percentage of the chemical component in the final product is in excess of 0.001% in leave on products and 0.01% in wash off products. Citral 0.08%, Geraniol 0.36%, Citronellol 0.83%.

Egyptian Rose Geranium may be used up to 100% of the fragrance concentrate at a usage rate of up to 5% of the total weight in the finished product under category 9.

Guaiacwood (*Bulnesia sarmienti*)

Guaiacwood oil, also known as champaca wood oil imparts a delicate tea rose odor when blended with other oils. In the bottle, a definite smoky odor can be detected along with the rose, but is not apparent in the final soap. Guaiacwood oil is used as a modifier and fixator and is one of the finest fixatives in soap perfumery. It is especially well suited for blending with florals, providing depth and sweetness to a blend. It's also used as a base in the production of rose and sandalwood fragrances.

IFRA Restrictions: None at this time.

IFRA Specifications: None at this time.

EU Cosmetic Directive, Annex III, 7th Amendment: None at this time.

Guaiacwood oil may be used up to 100% of the fragrance concentrate at a usage rate of up to 5%.

Lavandin Super (*Lavandula x intermedia*)

Lavandin is the hybrid cross of true lavender and spike lavender. True lavender has a delicate and floral lavender aroma while spike lavender comes across as more camphoraceous and penetrating in its representation. The two were crossed in order to gain the higher yield oil spike lavender produces yet maintain the typical lavender scent in the process. There are many lavandin hybrids available but lavandin super more closely resembles true lavender due to its higher proportion of linalyl acetate.

IFRA Restrictions: Lavandin oil contains the following chemical components restricted under IFRA guidelines: Coumarin < 0.3%, Geraniol < 0.6%, 1-Octen-3-yl acetate < 1.5%.

IFRA Specifications: Lavandin oil contains the following chemical components specified under IFRA guidelines: Linalol, limit peroxide levels to 20 mmol/l.

EU Cosmetic Directive, Annex III, 7th Amendment: Lavandin oil contains the following chemical components that must be listed on the label if the percentage of the chemical component in the final product is in excess of 0.001% in leave on products and 0.01% in wash off products. Coumarin < 0.3%, Geraniol < 0.6%.

Lavandin oil may be used up to 100% of the fragrance concentrate at a usage rate of up to 5% of the total weight in the finished product under category 9. Consideration must also be given to 1-Octen-3-yl acetate. If used in combination with other essential oils or fragrance components containing this ingredient, the combined percentage of the restricted ingredient in all fragrance components within the formula must be calculated at less than 0.3% in the finished product.

Lavender (*Lavandula angustifolia*)

The Roman's used lavender to scent their public baths. Mary washed and anointed the feet of Jesus with lavender. In Medieval Europe washerwomen were called lavenders for their habit of hanging the laundry to dry on lavender bushes and then placing the laundry in lavender scented drawers.

To the general public, lavender is synonymous with handcrafted soap and any soapmaker of merit understands the wisdom of having lavender in their product line. While it lends itself beautifully as a blender, sweetener and modifier with other oils, it is difficult to blend a true lavender smelling soap having any appreciable shelf life. It has a tendency to oxidize in soap and the scent fades quickly. Lavender must be strengthened and well fixed. Lavandin may be a better substitute however it too requires careful blending.

IFRA Restrictions: Lavender oil contains the following chemical components restricted under IFRA guidelines: Coumarin < 0.1%, Geraniol < 1%, 1-Octen-3-yl acetate < 1%.

IFRA Specifications Lavender oil contains the following chemical components specified under IFRA guidelines: Linalol, limit peroxide levels to 20 mmol/l.

EU Cosmetic Directive, Annex III, 7th Amendment: Lavender oil contains the following chemical components that must be listed on the label if the percentage of the chemical component in the final product is in excess of 0.001% in leave on products and 0.01% in wash off products. Coumarin < 0.1%, Geraniol < 1%.

Lavender oil may be used up to 100% of the fragrance concentrate at a usage rate of up to 5% of the total weight in the finished product under category 9. Consideration must also be given to 1-Octen-3-yl acetate. If used in combination with other essential oils or fragrance components containing this ingredient, the combined percentage of the restricted ingredient in all fragrance components within the formula must be calculated at less than 0.3% in the finished product.

Lemongrass (*Cymbopogon citratus*)

Historically, the ancient Greeks, Romans and Egyptians used lemongrass to relieve tired muscles, treat fevers and infections and in the manufacture of cosmetics and perfumes. Today it is used extensively in aromatherapy and personal care products for the treatment of acne, cellulite and sore muscles. It is used primarily in soap perfumery for its lasting lemon scent and as a basis for lemon verbena. Use geranium or citronella to modify any harsh grassy undertones.

IFRA Restrictions: Lemongrass oil contains the following chemical components restricted under IFRA guidelines: Citral < 82%, dl-Citronellol < 0.8%, Geraniol < 8%, Eugenol < 0.3%.

IFRA Specifications: None at this time.

EU Cosmetic Directive, Annex III, 7th Amendment: Lemongrass oil contains the following chemical components that must be listed on the label if the percentage of the chemical component in the final product is in excess of 0.001% in leave on products and 0.01% in wash off products. Citral < 82%, dl-Citronellol < 0.8%, Geraniol < 8%, Eugenol < 0.3%.

Lemongrass oil may be used up to 100% of the fragrance concentrate at a usage rate of up to 5% of the total weight in the finished product under category 9.

If other essential oils or fragrance components within the fragrance concentrate also contain eugenol, the combined percentage of all eugenol containing ingredients must not exceed 0.5% in the finished product.

Oakmoss (*Evernia prunastri*)

Oakmoss is a lichen found on oak trees. Other mosses found on spruce and pine trees are also called oakmoss in the USA, but are less refined. Most soapmakers employ the concrete, which is a solid or semi-solid mass. It has an extremely pleasant, deep and earthy aroma and is an excellent fixative and modifier in almost any blend. Although oakmoss appears to be too expensive for the handcrafted soapmaker, a little goes a long way. The odor is so tenacious it is advisable to dilute the oakmoss either with perfumers' alcohol or carrier oil by 75%. A 2-ounce (57 gram) bottle of oakmoss will make 8 ounces (227 grams) of tincture.

IFRA Restrictions: Oakmoss extracts are limited to 0.1% in the finished product.

IFRA Specifications: Oakmoss extracts used in fragrance compounds must not contain added tree moss, which is a source of resin acids. Traces of resin acids may be carried over to commercial qualities of oak moss in the manufacturing process. These traces must not exceed 0.1% dehydroabietic acid (DHA) in the extract.

EU Cosmetic Directive, Annex III, 7th Amendment: The INCI name of Oakmoss, *Evernia prunastri*, must be listed on the label if oakmoss extract is included in the formula.

Oakmoss extract is restricted to a maximum of 0.1% in the finished product under Category 9 and 10 of the IFRA Guidelines. This would equate to a fragrance blend concentrate containing up to 10% oakmoss extract with a usage rate of up to 1% of the fragrance concentrate in the total weight of the finished product. With a usage rate of up to 5% of the fragrance concentrate in the finished product, the amount of oakmoss extract would be reduced to a maximum of 2% in the fragrance concentrate.

Consideration must be given to the following IFRA restricted ingredients in oakmoss extract if used in combination with other essential oils or fragrance components also containing these ingredients. The combined percentage of the restricted ingredient in all fragrance components within the formula must be calculated. The combined ingredients must not exceed 0.1% dehydroabietic acid.

Palmarosa (*Cymbopogon martinii*)

Palmarosa oil was historically used as a fragrance oil as well as for its therapeutic and insect repellent properties. It is great for mature skin for its cell regeneration and sebum regulating properties. It has been used extensively as an adulterant to rose oil. Although it has a rose-like scent, it also has a weedy back note difficult to ignore and requires careful blending.

IFRA Restrictions: Palmarosa oil contains the following chemical components restricted under IFRA: Citral < 1%, Farnesol < 1.2%, Geraniol < 85%.

IFRA Specifications: None at this time.

EU Cosmetic Directive, Annex III, 7th Amendment: Palmarosa oil contains the following chemical components that must be listed on the label if the percentage of the chemical component in the final product is in excess of 0.001% in leave on products and 0.01% in wash off products. Citral, Farnesol, Geraniol, Limonene < 1%, Linalol < 4%.

Palmarosa may be used up to 100% of the fragrance concentrate at a usage rate of up to 5% of the total weight in the finished product under category 9.

Patchouli (*Pogostemon cablin*)

Patchouli has a powerful, tenacious odor with a musty, earthy smell. Its dominating and distinctive smell has become synonymous with the hippy movement in the 1970's due to its widespread use in masking unpleasant and illegal odors. It deserves a better reputation due to its wonderful skin care benefits. Patchouli's cell regenerating properties make it especially useful for dry, mature and wrinkled skin. Its musty odor gives body and character to a blend and is an excellent fixative. A little goes a long way and it should be used with prudence. Unlike most other oils, patchouli oil improves with age.

IFRA Restrictions: None at this time.

IFRA Specifications: None at this time.

EU Cosmetic Directive, Annex III, 7th Amendment: Patchouli oil contains the following chemical components that must be listed on the label if the percentage of the chemical component in the final product is in excess of 0.001% in leave on products and 0.01% in wash off products. Linalol < 1%, Limonene < 1%, Eugenol < 1%.

Patchouli oil may be used up to 100% of the fragrance concentrate at a usage rate of up to 5% of the total weight in the finished product under category 9.

Peppermint (*Mentha piperita*)

Peppermint oil is known for its cooling effect on the skin and is especially suited in treating fevers and hot flashes. It is also found to be beneficial in hair care ingredients for the treatment of dandruff. Because of its cooling effect on the skin, it is beneficial only in small quantities as a perfume material. Peppermint finds use in soap perfumery in its ability to sweeten and add character. It is especially useful in modifying and sweetening geranium.

IFRA Restrictions: Peppermint oil contains the following chemical components restricted under IFRA guidelines: Carvone 2%, Trans-2-Hexenal < 0.1%.

IFRA Specifications: None at this time.

EU Cosmetic Directive, Annex III, 7th Amendment: Peppermint oil contains the following chemical components that must be listed on the label if the percentage of the chemical component in the final product is in excess of 0.001% in leave on products and 0.01% in wash off products. Linalol < 1%, Limonene < 5%.

Due to traces of the chemical component trans-2-Hexenal, Peppermint oil is restricted to a maximum of 2% in the finished product under Category 9 and 10 of the IFRA Guidelines. This would equate to a fragrance blend concentrate containing up to 100% peppermint oil with a usage rate of up to 2% of the fragrance concentrate in the total weight of the finished product. With a usage rate of 5% of the fragrance concentrate in the finished product, the amount of peppermint oil would be reduced to a maximum of 40% in the fragrance concentrate. If other essential oils or fragrance components within the fragrance concentrate also contain trans-2-Hexanol, the combined percentage of all trans-2-Hexenal containing ingredients must not exceed 0.002% in the finished product.

Peru Balsam (*Myroxylon balsamum*)

Peru Balsam, with its rich vanilla-like scent is known for its ability to aid in the treatment of chapped skin, rashes, eczema, sores and wounds. It is used in perfumery as a fixative and to bring warmth and body to a blend.

IFRA Restrictions: Peru balsam oil contains the following chemical components restricted under IFRA guidelines: Coumarin < 0.1%, Isoeugenol < 0.2%, Benzyl alcohol < 5%, Benzaldehyde < 0.2%, Benzyl benzoate < 75%, Benzyl cinnamate < 9%.

IFRA Specifications: None at this time.

EU Cosmetic Directive, Annex III, 7th Amendment: Peru balsam oil contains the following chemical components that must be listed on the label if the percentage of the chemical component in the final product is in excess of 0.001% in leave on products and 0.01% in wash off products. Benzyl alcohol < 5%, Coumarin < 0.1%, Benzyl cinnamate < 9%, Benzyl benzoate < 75%.

The IFRA prohibits the use of crude Peru balsam as a fragrance ingredient; however Peru balsam extracts and distillates may be used as a fragrance ingredient of cosmetic products up to a maximum concentration of 0.4% in the finished product for Category 9 and 10. This would equate to a fragrance blend concentrate containing up to 40% Peru Balsam with a usage rate of up to 1% of the fragrance concentrate in the total weight of the finished product. With a usage rate of up to 5% of the fragrance concentrate in the finished product, the amount of Peru balsam oil would be reduced to a maximum of 8% in the fragrance concentrate.

Consideration must also be given to the following IFRA restricted ingredients in Peru balsam oil. If used in combination with other essential oils or fragrance components containing these ingredients, the combined percentage of the restricted ingredient in all fragrance components within the formula must be calculated. The IFRA restricted ingredients are listed showing the maximum percentage allowed in the finished product

Chemical Isolate	Maximum Percentage Allowed in Finished Product for IFRA Category 9 & 10
Benzaldehyde	3% Category 9 & 2.5% Category 10
Isoeugenol	0.02%

Rosemary (*Rosemarinus officinalis*)

Before the advent of antibiotics, rosemary oil was the number one defense against the spread of disease in hospitals. Rosemary was strewn in the hallways, burned and worn around the neck to keep the dreaded plague and other communicable diseases at bay. In the middle ages it was credited with magical powers strong enough to ward off evil spirits.

Widely recognized for its ability to aid in mental clarity, it is commonly referred to as the remembrance plant and is used as a sign of love and friendship. Rosemary oils fresh, camphoraceous odor blends well with all herbal and citrus oils. Its regenerating qualities make it especially beneficial to skin and hair care. In soap perfumery it has a special affinity for lavender and aids in strengthening its scent. Rosemary is very useful in many soap blends for adding lightness and freshness.

IFRA Restrictions: None at this time.

IFRA Specifications: None at this time.

EU Cosmetic Directive, Annex III, 7th Amendment: Rosemary oil contains the following chemical components that must be listed on the label if the percentage of the chemical component in the final product is in excess of 0.001% in leave on products and 0.01% in wash off products. Linalol 35%, D Limonene < 3%, Geraniol < 2%.

Rosemary oil may be used up to 100% of the fragrance concentrate at a usage rate of up to 5% of the total weight in the finished product under category 9.

Dalmatian Sage (*Salvia officinalis*)

Considered a sacred herb by the Romans, Dalmatian sage is known for its herbaceous warm and spicy pine like odor. Besides its use in the dressing of the Thanksgiving turkey, it finds its way in perfumery for its fresh, clean out doors scent. Although no IFRA restrictions have yet been placed on Dalmatian sage, it should be used with care due to its high thujone content.

IFRA Restrictions: None at this time.

IFRA Specifications: None at this time.

EU Cosmetic Directive, Annex III, 7th Amendment: Dalmatian sage oil contains the following chemical components that must be listed on the label if the percentage of the chemical component in the final product is in excess of 0.001% in leave on products and 0.01% in wash off products. Linalol < 0.6%, Limonene < 3%.

Dalmatian Sage may be used up to 100% of the fragrance concentrate at a usage rate of up to 5% of the total weight in the finished product under category 9.

Spearmint (*Mentha spicata*)

Historically spearmint has been used to treat headache, ringworm, sexually transmitted diseases, bad breath, wounds and scars. Milder on the skin than peppermint, it is used extensively in soap perfumery for its sweet refreshing scent.

IFRA Restrictions: Spearmint oil contains the following chemical components restricted under IFRA guidelines: Carvone: Chinese Spearmint 80 Oil < 80%, Native Scottish Spearmint Oil < 70%, Chinese Spearmint 60 Oil < 60%; Trans-2 Hexenal traces < 0.1%.

IFRA Specifications: Spearmint oil contains the following chemical components specified under IFRA guidelines: Carvone, limit peroxide levels to 20 mmol/l.

EU Cosmetic Directive, Annex III, 7th Amendment: Spearmint oil contains the following chemical components that must be listed on the label if the percentage of the chemical component in the final product is in excess of 0.001% in leave on products and 0.01% in wash off products. Limonene < 15.597%, Linalol < 2.106%.

Due to traces of the chemical component trans-2-Hexenal, spearmint oil is restricted to a maximum of 2% in the finished product under Category 9 and 10 of the IFRA Guidelines. This would equate to a fragrance blend concentrate containing up to 100% spearmint oil with a usage rate of up to 2% of the fragrance concentrate in the total weight of the finished product. With a usage rate of 5% of the fragrance concentrate in the finished product, the amount of spearmint oil would be reduced to a maximum of 40% in the fragrance concentrate. If other essential oils or fragrance components within the fragrance concentrate also contain trans-2-Hexanol, the combined percentage of all trans-2-Hexenal containing ingredients must not exceed 0.002% in the finished product.

Sweet Orange Oil Folded (*Citrus sinensis*)

Orange oil has recently become popular for its natural ability to clean and refresh. It is used in perfumery to sweeten almost any blend and is used as a main constituent in eu-de-colognes. Typically, pure orange oil makes a poor choice for the soap pot due to its low flashpoint and its tendency to oxidize. Folded orange oil on the other hand is stronger, richer and more stable due to the high aldehyde content and terpeneless nature of the oil. Choose folded orange oil of 10X or higher for optimum value.

IFRA Restrictions: Folded orange oil contains the following chemical components restricted under IFRA guidelines: p-Mentha-1,8-dien-7-al (Perilla aldehyde) < 0.3.

IFRA Specifications: Sweet orange oil contains the following chemical components specified under IFRA guidelines: D-Limonene, Linalol, limit peroxide levels to 20 mmol/l.

EU Cosmetic Directive, Annex III, 7th Amendment: Sweet orange oil contains the following chemical components that must be listed on the label if the percentage of the chemical component in the final product is in excess of 0.001% in leave on products and 0.01% in wash off products. D-Limonene < 91.92%, (5 fold), < 81.71% (10 fold), Linalol < 1.01% (5 fold), < 2.08% (10 fold).

Sweet orange oil folded may be used up to 100% of the fragrance concentrate at a usage rate of up to 5% of the total weight in the finished product under category 9.

If other essential oils or fragrance components within the fragrance concentrate also contain p-Mentha- 1,8-dien-7-al (Perilla aldehyde), the combined percentage of all p-Mentha- 1,8-dien-7-al (Perilla aldehyde) containing ingredients must not exceed 0.1% in the finished product.

Tea Tree (*Melaleuca alternifolia*)

Tea tree oil was first used by the Aboriginal people to treat all manner of infections from cuts and wounds to respiratory tract infections. It was given to Australian soldiers during WW II as a part of their first aid kits. Almost forgotten with the introduction of antibiotics, its ability to treat bacteria, fungus and viral infection has led to a resurgence of the oil. It is effective in a wide range of skin conditions including acne, dandruff, athlete's foot and herpes.

IFRA Restrictions: None at this time.

IFRA Specifications: None at this time.

EU Cosmetic Directive, Annex III, 7th Amendment: Tea tree oil contains the following chemical components that must be listed on the label if the percentage of the chemical component in the final product is in excess of 0.001% in leave on products and 0.01% in wash off products: Limonene, Linalol.

Tea tree oil may be used up to 100% of the fragrance concentrate at a usage rate of up to 5% of the total weight in the finished product under category 9.

Thyme (*Thymus vulgaris*)

Thyme has a sharp herbaceous odor with a warm and spicy undertone. It is excellent at adding strength and body to herbal and fresh type blends. It is used extensively in skin care for its antibacterial and astringent properties for treating eczema, acne, cuts and burns.

IFRA Restrictions: Thyme oil contains the following chemical components restricted under IFRA guidelines: Geraniol $\leq$ 1.5%, Citral $\leq$ 0.5%, Trans-2 Hexenal traces < 0.1%.

IFRA Specifications: None at this time.

EU Cosmetic Directive, Annex III, 7th Amendment: Thyme oil contains the following chemical components that must be listed on the label if the percentage of the chemical component in the final product is in excess of 0.001% in leave on products and 0.01% in wash off products. Geraniol, Limonene < 1%, Linalol < 5%, Citral.

Due to traces of the chemical component trans-2-Hexenal, Thyme oil is restricted to a maximum of 2% in the finished product under Category 9 and 10 of the IFRA Guidelines. This would equate to a fragrance blend concentrate containing up to 100% Thyme oil with a usage rate of up to 2% of the fragrance concentrate in the total weight of the finished product. With a usage rate of 5% of the fragrance concentrate in the finished product, the amount of thyme oil would be reduced to a maximum of 40% in the fragrance concentrate. If other essential oils or fragrance components within the fragrance concentrate also contain trans-2-Hexanol, the combined percentage of all trans-2-Hexenal containing ingredients must not exceed 0.002% in the finished product.

Ylang-Ylang II and III (*Cananga odorata*)

Ylang-Ylang is known as the "flower of flowers." Its heady scent is its main constituent but it is used extensively in hair rinses to encourage hair growth and is mildly effective in skin care as well. Ylang-Ylang, lends a beautiful note to powders and is much used in Oriental perfumes. Distillation of the oil is carried out over three distillations with the first being superior. Soapmakers typically use the second or third distillation referred to as Ylang-Ylang II or III. Considered a fixative in its own right, Ylang-Ylang lends a tenacious floral note to soap perfumery.

IFRA Restrictions: Ylang-Ylang oil contains the following chemical components restricted under IFRA: Benzyl alcohol < 0.5%, Benzyl benzoate < 8.5%–10%, Benzyl salicylate < 4%–5%, Eugenol < 0.5%, Farnesol < 3%–4%, Geraniol < 2.4–0.8%, Isoeugenol < 0.5%. Ylang-Ylang extracts are restricted to 5% of the finished product for Category 9 and 2.5% of the finished product for category 10 under the IFRA standards.

IFRA Specifications: None at this time.

EU Cosmetic Directive, Annex III, 7th Amendment: Ylang-Ylang oil contains the following chemical components that must be listed on the label if the percentage of the chemical component in the final product is in excess of 0.001% in leave on products and 0.01% in wash off products. Benzyl benzoate, Benzyl salicylate, Eugenol, Farnesol, Geraniol, Isoeugenol, Linalol < 10.167%.

Ylang-Ylang II and III may be used up to 100% of the fragrance concentrate at a usage rate of up to 5% of the total weight in the finished product under category 9.

Consideration must be given to the following IFRA restricted ingredients in Ylang-Ylang extracts if used in combination with other essential oils or fragrance components also containing these ingredients. The combined percentage of the restricted ingredient in all fragrance components within the formula must be calculated. The IFRA restricted ingredients are listed showing the maximum percentage allowed in the finished product.

Cananga odorata, commonly called Ylang-Ylang. Plate from Francisco Manuel Blanco's Flora de Filipinas, 1883.

Chemical Isolate	Maximum Percentage Allowed in Finished Product for IFRA Category 9 & 10
Eugenol	0.5%
Isoeugenol	0.02%

Blending Essential Oils for Scenting Soap Naturally

Imagine the smell of lemon, lavender and patchouli beautifully blended together without one being lost by the other.

The ability to smell an essential oil with the imagination as well as the nose and to blend accordingly is one of the keys to successful blending. The easiest way to achieve this is to work with the oils and learn about their character and unique personalities.

The goal of most soapmakers is to achieve a pleasant, lasting and cost effective scent blend. The internet is full of information on blending, but most begin and unfortunately end with a technique utilizing the philosophy of top note (light and airy), middle note (the heart and ultimate dry down note) and base note (the note that fixes the scent so it will last). The suggested ratio is 3-2-1. Three parts of the blend is the top note, two parts the middle note and 1 part the base note making up the total. In a perfect world, the recipe for the lemon, lavender and patchouli blend would work like this:

> 3 parts lemon = 50%
> 2 parts lavender = 33%
> 1 part patchouli = 17%

One of the problems encountered with this method is working in parts. Parts make the math more difficult when trying to come to a precise measurement. Working in percentages is much easier. The biggest problem with this technique is it is limiting in its simplicity and does not take into account the individuality of the essential oils or the "soap note." Soap, especially liquid soap, has a "note" of its own. The "soap note." is an ever present obstacle to the soapmaker if a fragrance is not carefully blended. A poorly blended fragrance can oxidize the top note, mutate the middle note and leave a lingering scent of the base note with an overriding, somewhat fatty back note in its wake.

In blending scents for liquid soap, it is necessary to know the strength of the oils and how they behave in contact with the soap so well-informed decisions may be made regarding the proportion as well as the type of oils that will create a successful blend. Most blending knowledge is gained through experience, but the soap-maker can get a head start by understanding the following tips:

- Oils with a higher flashpoint will generally have more staying power than oils with a lower flashpoint. Exceptions are camphoraceous oils such as rosemary, lavandin, eucalyptus, peppermint and thyme. Although their flashpoint is lower, the camphoraceous note of the oil lends it strength and tenacity.

- Oils with a higher specific gravity will generally have more staying power than oils with a lower specific gravity.

- Citrus oils and other oils high in terpenes have a tendency to oxidize in liquid soap by either fading out or lending an off odor (think paint thinner) to the soap. Lower percentages of these oils may be used if the blend is well supported by other oils. Folded citrus oils in which much of the terpenes have been removed are better alternatives.

- Oils with a high percentage of esters, such as linalyl acetate in lavender oil, should be well fixed to prevent deterioration.

- Oils distilled from grasses, such as citronella, lemongrass, palmarosa and patchouli are more tenacious than oils distilled from the leaves of plants.

- The further out on the limb, the more fleeting the scent. Oils distilled from the fruit, such as lemon oil, are more volatile than the oil distilled from the flower of the same tree. Oils distilled from the wood, such as cedar wood, are more stable than oils distilled from the berry or leaf.

Looking again at the lemon, lavender and patchouli blend, how will it hold up in a soap formula?

Lemon oil = citrus = fruit = high terpenes = low flashpoint = low specific gravity= very poor performance.

Lavender oil = flower = high ester content =medium flashpoint = low specific gravity = poor performance.

Patchouli oil = grass = high flashpoint = high specific gravity = excellent performance.

While some oils will easily fit into a 3-2-1 blend, most won't as in the case of the above example. In order to achieve the desired effect it is necessary to "build" the blend with better choices of essential oils.

When building, it is necessary to begin with a **base**. The **Base** (not to be confused with base notes) is the essence of the blend, the primary essential oils that will evoke the combination of top, middle and base notes to create the desired aroma. Upon the base it is sometimes necessary to add **Blenders, Modifiers, Sweeteners and Fixators.**

- **Blenders** harmonize with the essential oils in the base. They can be used to add strength, reduce cost, or support a more heavily regulated essential oil to help achieve the desired odor.

- **Modifiers** may be used to minimally alter the blend by giving it a lighter or darker note or by adding a touch of mystery.

- **Sweeteners** add a more pleasing effect.

- **Fixators** or fixatives as they are commonly called decrease the flashpoint and give the blend strength and tenacity. They also provide the blend with warmth and body.

Building a simple blend with a balanced odor of lemon, lavender and patchouli could therefore be based on the following choices:

- **Base: Considerations for lavender include:** Lavander, lavandin, spike lavender. Lavandin is a better choice than lavender because it holds up better in soap.

  Considerations for lemon include: Lemon, lemongrass citronella, lemon eucalyptus, litsea cubeba. Lemongrass has the most tenacious lemon scent over lemon and litsea cubeba and tenacity is especially important when paired with patchouli. Lemon eucalyptus and citronella both have undertones that would be less desirable, but could be used to blend with the lemongrass.

  Patchouli: Patchouli stands alone with its unique scent.

  The base would therefore be comprised of lavandin (top note) lemongrass (middle note) and patchouli (base note).

- **Blenders:** Lavender, spike lavender, citronella, lemon eucalyptus and litsea cubeba could all be used to blend with the lavandin and lemongrass.

- **Modifiers:** Rosemary, spearmint, thyme, clary sage, bergamot and folded orange oils blend beautifully with both lavandin and lemongrass lending strength, lightness and depth to the blend.

- **Sweeteners:** Spearmint, folded orange oil, geranium and bergamot would add sweetness.

- **Fixators:** Patchouli alone or in combination with any essential oils with a high flashpoint that blend well with patchouli including guaiacwood, cedar, clove bud and sandalwood.

Some essential oils as well as fragrance oils may cloud or thicken liquid soap. Almost all cloud when mixed with room temperature soap but will clear upon standing. Adding the fragrance to slightly warm soap will help to avoid this step. If the soap remains cloudy after cooling, using specially denatured alcohol or perfumers' alcohol in the scent blend at a rate of 50%–100% of the weight of the concentrated blend will eliminate clouding in most all fragrances. Polysorbate 20, an emulsifier, will help prevent stubborn fragrance and essential oils from separating out of the soap. Use it up to 100% of the concentrated scent blend as an emulsifier to alleviate separation, clouding and thickening of the soap.

Liquid soap is adequately scented at a rate of 0.5–1% of the finished weight of the soap. All recipes given below are within the IFRA standards at that percentage of use. Recipes with an asterisk will not meet IFRA standards at a 5% usage rate and will need to be reformulated for compliance. Reformulating can be made easier by the use of a fragrance allergen calculator that can easily be found online.

Therefore, starting with a simple blend based on the example above and utilizing the essential oils with IFRA guidelines as given, the following recipes are presented.

Scent Blends Designed for Liquid Soap

LAVENDER, LEMONGRASS & PATCHOULI

This is a simple formula based on the blending example above.

Lavandin	55% – base
Lemongrass	27% – base
Spearmint	10% – modifier/sweetener
Patchouli	8% – base/fixator

LAVENDER, LEMONGRASS & PATCHOULI

A more complicated, full bodied blend also based on the above blending example.

Lavender	15% – base
Lavandin	27% – blender
Lemongrass	15% – base
Lemon Eucalyptus	5% – blender
Rosemary	10% – modifier
Bergamot	8% – modifier/sweetener
Spearmint	10% – modifier/sweetener
Patchouli	7% – base/fixator
Guaiacwood	3% – fixator

Lemongrass

The most popular scents found in hotel bath and body amenities include lemon or citrus, most likely due to their clean fresh crispness. This is one of Gaily Rebecca's favorites.

Lemongrass	58%
Bergamot	10%
Lavandin	8%
Rosemary	5%
Clove	6%
Folded Orange	4%
Spearmint	3%
Geranium	2%
Guaiacwood	4%

Citrus Splash

A balanced blend of citrus and sweet!

Lemongrass	29%
Folded Orange	27%
Bergamot	20%
Geranium	8%
Lavandin	7%
Clove	3%
Guaiacwood	6%

LEMON TEA

An excellent antioxidant blend!

Bergamot	35%
Lemongrass	20%
Rosemary	20%
Lavandin	10%
Ylang-Ylang	8%
Peru Balsam	5%
Oakmoss	2% (25% tincture of oakmoss in perfumers' alcohol or carrier oil)

CAMPER'S COMPANION

A refreshing blend of oils recognized for their ability to keep bugs away!

Lemon Eucalyptus	23%
Lemongrass	10%
Lavandin	20%
Spearmint	15%
Rosemary	15%
Cedarwood	17%

Sweet Orange

A well fixed orange with floral notes and modifiers to give it a fresh "kick!"

Folded Orange	55%
Bergamot	15%
Geranium	10%
Lemon Eucalyptus	10%
Guaiacwood	10%

Orange Spice*

This is a nice warm citrus scent for a cold winter day. Due to the spice oils, usage rate is limited to a maximum of 3% in the final product due to the cassia oil. To use at 5% maximum, reduce the cassia to 1%.

Folded Orange	52%
Clove	15%
Bergamot	15%
Geranium	10%
Cassia	1.5%
Guaiacwood	6.5%

Lavender I*

*Because of the necessity of "fixing" lavender in a soap blend, there are many interpretation of this all-time favorite. This is one of them. *If using at concentrations higher than 4%, the Peru balsam should be recalculated to meet IFRA standards.*

Lavandin	50%
Bergamot	15%

Peru Balsam	10%
Guaiacwood	10%
Clary Sage	8%
Spearmint	5%
Geranium	2%

LAVENDER II*

This is a nice blend of lavender even the guys will love. *If using at concentrations higher than 3%, recalculate the Peru balsam to IFRA standards.

Lavandin	57%
Peru Balsam	13%
Bergamot	12%
Clary Sage	8%
Guaiacwood	8%
Oakmoss	2% (25% tincture of oakmoss in perfumers' alcohol or carrier oil)

LAVENDER & PATCHOULI*

A lavender that is sure to please patchouli lovers. If using at a concentration higher than 2% recalculate the Peru balsam to IFRA standards.

Lavandin	70%
Peru Balsam	17%
Patchouli	5%
Clove	5%
Ylang-Ylang	2%
Geranium	1%

ROSEMARY & LAVENDER I

Rosemary and lavender are one of the most popular scents of the handcrafted soapmaker. This is a fresh & vibrant version.

Lavandin	35%
Rosemary	25%
Geranium	13%
Thyme	8%
Guaiacwood	8%
Spearmint	5%

ROSEMARY & LAVENDER II

This is a darker, fuller bodied rosemary and lavender blend.

Lavandin	45%
Rosemary	18%
Lavender	15%
Bergamot	8%
Geranium	4%
Peru Balsam	3%
Thyme	2%
Oakmoss	2% (25% tincture of oakmoss in perfumers' alcohol or carrier oil)

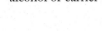

WHITE WINDSOR

An age old blend similar to rosemary and lavender.

Lavandin	36%
Bergamot	31%

Geranium	15%
Thyme	10%
Clove	8%

HERB*

An old fashioned herbal blend with a fresh uplifting scent!
*If using at concentrations higher than 1% in the finished product recalculate the cassia oil to IFRA standards.

Geranium	32%
Lavandin	23%
Rosemary	11%
Thyme	10%
Spearmint	7%
Clary Sage	6%
Lemon Eucalyptus	6%
Cassia	5%

ROSEMARY MINT

A very clean and energizing scent.

Rosemary	45%
Spearmint	30%
Peppermint	10%
Lavandin	10%
Guaiacwood	5%

Spearmint

A simple and fresh spearmint blend!

Spearmint	40%
Lavandin	39%
Rosemary	15%
Guaiacwood	6%

Peppermint

Same song second verse . . . .

Peppermint	40%
Lavandin	39%
Rosemary	15%
Guaiacwood	6%

Peppermint Tea Tree*

*A great blend for blemishes, dandruff and oily skin! Watch out around the undercarriage as peppermint is "cooling" all over! *If using at concentrations higher than 1% in the finished product, recalculate the cassia to IFRA standards.*

Peppermint	40%
Tea Tree	20%
Rosemary	15%
Geranium	15%
Cassia	3%
Cloves	7%

HAPPILY EVER AFTER*

*A nice clean spicy floral scent! *Do not use in concentrations higher than 1% in the finished formula due to the high concentration of clove in the formula.*

Geranium	50%
Clove	30%
Peppermint	20%

SEA BREEZE

A very fresh and invigorating blend to get going in the morning.

Rosemary	33%
Bergamot	20%
Lavender	17%
Folded Orange	13%
Peppermint	13%
Guaiacwood	4%

HERBAL FOREST*

*A 3-2-1 blend that works! *Do not use in concentrations higher than 1% in the finished formula due to the high concentration of Peru balsam in the formula.*

Lavandin	50%
Peru Balsam	33%
Patchouli	17%

Alpine Meadows*

*A fresh invigorating outdoor blend that is good for spring & summer scents. *If using at concentrations higher than 1% in the finished product recalculate the cassia oil to IFRA standards.*

Lavandin	19%
Peppermint	18%
Rosemary	16%
Sage	15%
Lemongrass	12%
Lemon Eucalyptus	10%
Cassia	5%
Guaiacwood	5%

Cedarwood Sage

A well rounded outdoorsy scent!

Atlas Cedarwood	30%
Dalmatian Sage	20%
Rosemary	20%
Lavandin	15%
Clove Bud	10%
Lemon Eucalyptus	5%

Woodlands*

*A spicy woody blend sure to please the guys! *If using at concentrations higher than 1% in the finished product recalculate the cassia oil to IFRA standards.*

Atlas Cedarwood	52%
Geranium	13%
Clove Bud	5%
Cassia	5%
Patchouli	4%
Ylang-Ylang	4%
Bergamot	4%
Oakmoss	13% (25% tincture of oakmoss in perfumers' alcohol or carrier oil)

Emollient

A mellow blend of oils known for their therapeutic benefits in treating all manner of skin complaints.

Atlas Cedarwood	60%
Lemongrass	32%
Palmarosa	8%

Down to Earth

A deep and rich full bodied blend.

Bergamot	52%
Geranium	23%

Atlas Cedarwood	16%
Patchouli	9%

FLOWER

A light sweet blend of floral notes hinting at rose.

Lavandin	30%
Palmarosa	22%
Bergamot	12%
Peru Balsam	8%
Clary Sage	8%
Guaiacwood	8%
Spearmint	5%
Lemon Eucalyptus	5%
Oakmoss	2% (25% tincture of oakmoss in perfumers' alcohol or carrier oil)

SPRING BOUQUET

A rich floral blend.

Geranium	24%
Lavender	20%
Ylang-Ylang	14%
Clove	11%
Bergamot	10%
Rosemary	10%
Peru Balsam	8%
Peppermint	3%

SWEET SERENITY*

*A deep, spicy and sweet floral. *If using at concentrations higher than 1% in the finished product recalculate the cassia oil to IFRA standards.*

Ylang-Ylang	50%
Lavender	19%
Folded Orange	10%
Peru Balsam	8%
Clove Bud	5%
Cassia	4%
Geranium	4%

CHAPTER

INCORPORATING ADDITIVES

Additives

"If there's one basic liquid soapmaking principle that needs to be underlined, emphasized, and shouted from the rooftops, it's this: Unneutralized fatty acids cause cloudiness."

~ *Making Natural Liquid Soaps*,
(Failor, 2000.)

It is in the handcrafted soapmaker's very nature to manipulate and coerce the proverbial square peg into the round hole, such is the story of how cold process soap, once touted as an inferior process that created an inferior product, has come to be recognized as some of the most stunning and luxurious soap on the market today. At first glance, liquid soap appears to lack the creative canvas cold process soap so readily provides. Because of its liquid nature, it can't be swirled, molded, layered or imbedded and until recently botanical extracts were thought to be an impossible contribution to the formula. To the handcrafted soapmaker, boring is unacceptable and in order to get the soapmaker's attention, ways to express the individuality, creativity and beauty of handcrafted soap is imperative. The use of thickeners, colorants and fragrances add interest and texture but the key to an exceptional soap is the nutrient rich botanicals and other additives handcrafted soapmakers pride themselves in incorporating into their soap.

Superfatting and Lye Discounts

Superfatting, (adding more oil to the formula than required to react to the lye and make soap) or discounting the amount of lye in the formula so the soap will have excess oils left in the formula is the go-to method for cold process soap and is very tempting to the liquid soapmaker. Why get so hung up on clarity when excess oils add nourishing, skin friendly benefits and totally eliminate the whole process of neutralizing excess alkali in the formula? Problems of oil separation can be solved through the use of Polysorbate 80 to emulsify the oils into the soap and prevent the milky film of oil from floating on the surface of an otherwise clear formula. Thickening with a cellulose type thickener will help prevent the free fatty acids from forming sludge at the bottom of the container. It would appear the square peg is sitting pretty in the round hole with all the kinks of superfatting liquid soap resolved, or are they?

What cannot be avoided in a superfatted liquid soap is the degradation of the soap by oxidation of the excess oils. Excess oils in a liquid soap formula will oxidize and turn rancid much more rapidly than in bar soap due to the high percentage of water in a liquid soap formula. The most important reason to avoid superfatting is the shelf life and stability of the soap. Transparency and sparkle is the signature of a well-made soap and a goal all soapmakers should strive to achieve.

Sequestering Agents

Sequestering agents add clarity and sparkle to the soap by binding with the minerals within the soapmaking ingredients and keeping them in suspension. Glycerin and potassium carbonate which have already been mentioned are excellent sequestering agents and are already included in large portion in the no paste and gel soap recipes. Adding either of these ingredients in the soap paste method of making soap is an option to increase the clarity of the soap. Add the potassium carbonate at a rate of up to 5% of the oils in the formula to the lye portion of the water. The glycerin can be added at any time during the soapmaking process or as a part of the dilution of the soap paste at a rate of up to 33% of the soapmaking oils. Make sure to discount the water by the amount of glycerin being added.

Another excellent choice is sorbitol.

Sorbitol is generally found naturally in various fruits and plants and produced from corn syrup. It has a pH of 6–7 and is highly soluble up to an 83% solution. It is non-irritating and has been affirmed as GRAS (Generally Recognized as Safe) by the U.S. Food and Drug Administration and is approved for use by the European Union and numerous countries around the world, including Australia, Canada and Japan. It is also an allowable ingredient in EcoCert (an organic certification organization). Sorbitol has similar characteristics to glycerin as a humectant and conditioning agent with the exception it neither loses nor absorbs appreciable water. In liquid and transparent soaps, sorbitol will add clarity and brilliance to the soap to a greater degree than sugar or glycerin. It also acts as a foam booster with good wetting properties and as a sequestrant, it will chelate iron, copper and aluminum ions in the water with good anti-redeposition properties especially if used in conjunction with gluconic acid. General usage rate is less than 10%. Too much may cause a dampening in the lather.

Alternative Liquids

Liquid soap generally contains up to 60%–75% water and the rest is soap mixed with a small percentage of other additives as previously discussed. The first thing the handcrafted soapmaker sees when looking at all the water is an opportunity to add value and nutrition. The question immediately arises, what can be used in place of the water and how to go about the substitution?

When considering what substitutions can be made in the place of water, it is critical to have an understanding of how the additive will react when mixed with lye before any attempt to add it to the soapmaking formula. Adding highly acidic liquids to a lye solution can create a powerful and dangerous reaction. Know the pH of the ingredient to be added. Never add ingredients with a low pH directly to the soap solution. Dilute them first. The stronger the acid, the higher should be the dilution rate. Honey should only be added to a finished soap as it creates quite a nasty reaction at any other step in the process. If unsure how an ingredient will react, it would be wise to test it first. Use a very weak (highly diluted with water) lye solution and place a drop of the ingredient (also diluted) to be tested in the solution and gently swirl. Look for an increase in heat or a change in color or aroma. If there is no reaction, continue slowly placing drops in the solution, stopping immediately at the first sign of any change. If very little or no reaction has occurred after a significant amount of the ingredient has been introduced into the

> Mixing alkalis with various sugars may produce carbon monoxide, a poisonous gas.

lye solution, slowly increase the strength of the lye solution and continue watching for any undesirable reaction. If little or no reaction is shown and the soapmaker feels confident the additive can be safely added to the solution, proceed slowly and cautiously.

When to add alternative liquids to liquid soap: The best time to add alternative liquids to a liquid soap formula is just before the soap reaches full trace. Reasons for adding at this time are:

1. Most liquid additives have a tendency to slow the soapmaking process (with the exception of solvent type additives such as alcohol or sugar). When adding the additive at light trace, saponification has already begun and the soapmaker will avoid prolonged stick-blending times. Cook times however, will still be longer.
2. Adding mildly acidic ingredients at light trace with soap stock that is partially saponified creates a milder reaction than adding at the beginning of the process.
3. Adding alternative liquids at light trace rather than during dilution gives them the opportunity to become a part of the formula. The finished soap will have a greater chance of being transparent rather than cloudy and the alternative liquid will be less likely to separate out of the soap.

The chart below shows a sample of alternative liquids commonly added to handcrafted soap along with their pH. Milk and milk type products are addressed separately.

Alternative Liquid	pH
Aloe Vera Juice	4–6
Cucumber Juice	5.12–5.78
Tea (strong tea will be more acidic)	4.9–6
Coffee (strong coffee will be more acidic)	4.1–5.0
Wine	2.9–4.2
Beer	3.7–4.1

As seen in the chart, most of the alternative liquids are acidic. It is necessary to adjust the pH of the liquid either through adjustments in the amount of potassium hydroxide in the formula or with potassium carbonate to prevent the acid in the liquid from neutralizing the potassium hydroxide in the soap formula. How much depends upon the pH as well as the percentage of the liquid additive in the formula and can be calculated using the titration methods as outlined on page 35.

Milk Soaps: Milk is the most common alternative liquid used by the handcrafted soapmaker in cold process soaps. It is even more problematic incorporating into liquid soap than cold process soap. The protein and sugar in the milk react with the lye, burning the sugars and breaking down the protein giving a distinct ammonia smell to the brew. The cold process soapmaker, by lowering the heat of the lye solution and oils avoids most of these unwanted side effects. The liquid soapmaker does not have the luxury of working with such lowered temperatures. Liquid soapmaking requires sustained heat in order to fully saponify the oils.

> Experimentations of several methods of incorporating milk using room temperature oils and lye to prevent darkening have continually failed to achieve full saponification of the oils, leaving a cloudy and over-alkaline soap paste regardless of how long the paste is allowed to stand. A modified version of Catherine Failors' alcohol lye method of making liquid soap, using room temperature oils has met with a modicum of success, but the amount of alcohol remaining in the formula negates the benefits of the added milk.

The addition of milk, whether goat milk, butter milk, skim milk or coconut milk will affect the color and smell of the soap. How much depends on the percentage and the type of milk employed. Coconut milk will not darken as deeply as goat milk but the color of milk soap will range from deep amber to black coffee. In its own way, it makes a beautiful soap and the odor subsides to a dark, nutty and not unpleasant odor that blends well with heavier fragrances such as vanilla, amber or patchouli.

If incorporating milk into liquid soap, adjustments may need to be made to account for the pH as well as the fat in the milk. The pH of milk varies between varieties but most are slightly acidic with a pH range of 6–7. When milk begins to turn sour however, the pH becomes more acidic, such as buttermilk, sour cream and yogurt with pH ranges of 4.4–5. The milk fat may also need to be considered when calculating the lye in the soap formula depending upon the amount and type of milk being employed. Making a liquid soap with 3% of a soap paste formula employing fresh skim milk needs no adjustment, but making a liquid soap using 3% heavy cream is a different story. Below is a table showing different types of milk with percentages and saponification values of the milk fats.

Type of Milk	Percentage of Milk Fat	Saponification Value
Heavy Whipping Cream	36%	210–233
Half and Half	10.5–18%	210–233
Whole Milk	3.25%–4%	210–233
Goat Milk	3%–6%	223.36–233.84
Coconut Milk	17%	248–265
Soy Milk	2%	189–195
Rice Milk	0.4%	181–189

An example of adjusting a formula using 3% heavy cream in a soap paste formula is below:

Soap Paste Formula without Milk:

60% Coconut
30% Olive
10% Castor

21.6 oz. (612 g) Coconut Oil
10.8 oz. (306 g) Olive Oil
3.6 oz. (102 g) Castor Oil

9 oz. (255 g) Potassium Hydroxide
23 oz. (652 g) Water

Total Weight of Soap Paste = 68 oz. (1,928 g)

Soap Paste Formula with Heavy Cream:

1. Calculate the amount of heavy cream necessary for the recipe: [68 oz. (1,928 g) total wt. paste × 0.03 (3%) = 2.04 oz. (58 g) cream]

2. Subtract the amount of heavy cream from the water portion of the recipe.
   [23 oz. (652 g) water − 2.04 oz. (58 g) cream = 20.96 oz. (593 g) water]

3. Calculate the amount of fat in the heavy cream using the percentages listed in the chart above.
   [0.36 (36%) fat × 2.04 oz. (58 g) cream = 0.73 oz. (21 g) fat]

4. Calculate the amount of potassium hydroxide required to saponify the additional fat (refer to "Chapter 15: How to Formulate Liquid Soap Recipes").
   [0.73 oz. (21 g) fat × 0.2215 (saponification value of milk fat) = 0.16 oz. (5 g) potassium hydroxide]

5. Add the additional potassium hydroxide required to the original recipe.
   [9 oz. potassium hydroxide + 0.16 oz. (5 g) additional potassium hydroxide = 9.16 oz. (260 g) potassium hydroxide total.] (If using sour cream, you would round this number up to 0.2 to offset the acidic nature of the sour cream).

The Adjusted Recipe Containing Heavy Cream:

60% Coconut

30% Olive

10% Castor

21.6 oz. (613 g) Coconut Oil

10.8 oz. (306 g) Olive Oil

3.6 oz. (102 g) Castor Oil

9.16 oz. (260 g) Potassium Hydroxide

20.96 oz. (594 g) Water

2.04 oz. (58 g) Heavy Cream

Total Weight of Soap Paste = 68.16 oz (1,932 g)

Extracts

Adding dried botanicals to the soap creates a messy sludge at the bottom of the bottle, but extracts of botanicals alone or in combination create all manner of nutrient rich interest to the soapmaking formula. Huge inroads have been made in regards to the availability of ingredients to the handcrafted soapmaker since *Making Liquid Soap* (Failor, 2000) was first published. Botanical extracts such as liquid silk, green tea, mallow, oat and nettle as well as a host of others are now available in ready to use concentrated solutions, which can be added directly to the finished product to enrich the formula. Look for extracts that are water soluble rather than oil soluble to prevent superfatting and clouding. Most extracts can be added directly to the finished soap without affecting clarity, but testing of a small sample first before compromising the whole batch is recommended.

Because ready to use botanical extracts can be expensive, many soapmakers opt to make their own infusions and tinctures. Water infusions have a strong tendency to turn the soap formula brown or muddy green, but oil infusions generally have minimal effect on the color and clarity of the soap. Oil infusions contribute oil soluble antioxidants such as carotene and vitamins A and E along with nutrient rich flavonoids, polyphenols, phytosterols and other bioactive ingredients. Simple and cost efficient, dried herbs can be slowly steeped in the oil without heat for a period of two or more weeks or they can be quickly infused into the oil by the use of low heat. Being a rather impatient soapmaker, the quick method is outlined below. A quick internet search will further guide the patient soapmaker on other extraction methods.

Making Botanical Oil Infusions

Step 1 Select the plant to be infused. Aromatic plants such as chamomile, lavender, rosemary, sage, peppermint, juniper and thyme will lend their scents to the oil and ultimately the soap. Plants with a high level of fat soluble vitamins, antioxidants and saponins such as calendula, St. John's wort, plantain and comfrey are also good choices.

Step 2 If the plant is picked fresh, it should be dried a day or two to reduce the moisture in the plant. Moisture increases the potential of mold and bacteria growth and will also add water to the oil which may throw off the lye calculations.

Step 3 Place the herbs in a crockpot or double boiler and cover the herb completely with the soapmaker's choice of oil, leaving at least an inch or two of oil above the herbs. This oil will be a part of the oils in the soap formula and can be anywhere from 10% to 100% of the soapmaking oil. Some oil will be absorbed by the herb and lost so allowances will need to be made to accommodate the loss.

Step 4 Using the lowest setting on the crockpot, or a gentle simmer of the water in the bottom of a double boiler, heat for two to six hours; ideal temperature is 100°–140°F (38°–60°C).

Step 5 Strain the oil using cheesecloth, coffee filter, paper towel or clean lint free cloth. To ensure the oil is completely free of botanical debris, a second straining may be necessary.

Step 6 An antioxidant such as ROE or vitamin E should be added at this time if it has not already been added. The oil is ready to be weighed and added into the soapmaking formula or bottled using a dark, sterile container and properly labeled for future use. Store infused oils in a cool dry area for optimum shelf life.

Make sure the infused oil is used to replace the same oil called for in the recipe. For instance, if infusing the botanicals in olive oil, the infused oil would replace an equal amount of the olive oil in the soapmaking formula. If infusing the botanicals in coconut oil, it would replace the corresponding amount of the coconut oil in the formula.

Step 7 Proceed as usual with the preferred soapmaking method. Infused oils will work in any of the soapmaking methods and formulas provided.

CHAPTER

SHELF LIFE AND STABILITY

"Oxidation in air (auto-oxidation) can destroy fragrances and cause unpleasant odors from the breakdown or reaction of different ingredients with oxygen. The presence of unsaturation intensifies this reaction."

~ PRESERVATIVES FOR COSMETICS (STEINBERG, 2006)

The shelf life and stability of a product is dependent foremost upon the principles of good manufacturing practices and the selection of quality fresh ingredients. Attempting to preserve a product with ingredients compromised by poor hygiene or imperfect storage will produce an unreliable and inferior product. The goal of antioxidants, chelating agents and preservatives is to protect the product from decomposition over a reasonable shelf life and throughout its use by the consumer, not to compensate for the inefficiencies of the manufacturer.

Beyond good manufacturing practices lie other problems that affect the shelf life and stability of the product. These issues can be addressed through the choices of the ingredients that go into the soap pot, improved soapmaking techniques and due diligence on the part of the soapmaker.

Rancidity

Rancidity causes off odors, off colors and instability within liquid soap. It makes its presence known by the weakening of the fragrance before asserting its typical rancid smell and darkening the golden hue of the soap. It then continues past off odor and off color to instability as free fatty acids precipitate out of solution and become a white sludge at the bottom of the container.

All natural oils will become rancid, some sooner than others. The soapmaker can postpone rancidity by understanding and avoiding the causes.

To understand rancidity, it is important to know why it happens. There are two causes in liquid soap, oxidative rancidity and microbial rancidity.

Oxidative Rancidity

Oxidative rancidity is caused by the reaction of the oxygen in the air with the fat or the oil. Heat, light and trace metals increase the rate of this reaction.

Any soapmaker will understand the vulnerability of the oils during the soapmaking process and the soap during its shelf life. Adopting some of the following strategies can improve the odds.

Choose Oils Wisely: Choosing the right oils is the first essential element in the battle against oxidation.

Polyunsaturated oils such as corn, soybean and flaxseed are at the greatest risk of oxidation because of their high percentage of linoleic and linolenic fatty acids. Linoleic and linolenic fatty acids have more than one double bond in their structure. Like the dirt that collects in the corners and crevices of the kitchen floor, double bonds are the weakest part of the link that kinks the chain and leaves the oil vulnerable to attack by oxygen-free radicals.

Unsaturated oils such as olive, avocado and almond are less susceptible to oxidation than polyunsaturated oils because the majority of the fatty acids are comprised of oleic acid, which has only one double bond on its monounsaturated fatty acid chain.

Saturated oils such as palm, tallow and coconut are most stable against rancidity because of their high percentage of palmitic, stearic and/or myristic fatty acids, which are straight chain fatty acids that do not contain vulnerable double bonds.

Use Proper Manufacturing and Storage Procedures: Proper manufacturing techniques and storage impact the shelf life and stability of the soap as well.

The soapmaker should:

- Avoid unnecessarily high temperatures during the melting phase.

- Guard against overcooking the soap.

- Avoid over-evaporation of the soap paste.

- Keep the soap covered as much as possible throughout the procedure.

- Ensure the soap is neutral and free from unsaponified oils and free fatty acids.

- Store the finished soap or soap paste in a container with as little air space remaining as possible.

- Store the finished soap at an ambient temperature out of direct light.

Consider the Use of Antioxidants: The best defense against rancidity is the addition of antioxidants to the oils at the first opportunity. Their purpose is to protect the oils and fats from oxidation by free radicals (unstable oxygen molecules looking to steal electrons from the fat or oil).

Antioxidants are not preservatives. They do not protect the product from microbial contamination, nor do they reverse oxidation. Antioxidants have little or no effect if the oils are already compromised.

It is better to err on the side of caution when adding antioxidants to oils. Adding them in excess will have the opposite effect by creating a condition known as pro-oxidation, which actually promotes oxidation by free radicals. Antioxidants suitable for liquid soap, including a brief description of each are listed below in order of most synthetic to most natural.

> **BHA** also known as butylated hydroxyanisole is a chemical derivative of phenol. It is considered safe as a cosmetic ingredient in the present practice of use but has been listed as a substance known to cause cancer on California's Proposition 65 list. Usage rate is typically 0.02%. It is soluble in propylene glycol, fats and oils.
>
> **BHT** also known as butylated hydroxytoluene is also a chemical derivative of phenol. It is considered safe as used in cosmetic formulations. Usage rate is typically 0.01% to 0.1% by weight. It is soluble in fats and oils.
>
> **Propyl Gallate** is an ester formed by the condensation of gallic acid and propanol. It is twice as effective as BHA and BHT. In the United States it is considered generally recognized as safe (GRAS). Propyl gallate may react with metals such as iron and copper and discolor. Recommended usage rate is less than 0.1% by weight. It is somewhat soluble in water, is soluble in alcohol and propylene glycol with negligible solubility in fats and oils.

TBHQ also known as tert-butylhhdroquinone is a type of phenol and a derivative of hydroquinone. It is substantially more effective than BHA, BHT and propyl gallate. Typical usage is 0.01% and it is considered safe at usage rates up to 0.1%. It is slightly soluble in water and soluble in propylene glycol, fats and oils.

Tocopherols also known as vitamin E can be derived naturally from fats, oils and their plant sources or synthetically. In determining whether or not the specific tocopherol is natural or synthetic, it is important to note the difference in reference. Synthetic tocopherols are referred to as dl-tocopherols while natural tocopherols are referred to as d-tocopherols.

This discussion will focus on the natural d-tocopherol blend of alpha, gamma and delta most available to the soapmaker. This natural blend has greater antioxidant properties than its synthetic counterpart dl tocopherol due to the addition of the more active antioxidant qualities of the gamma and delta tocopherol. Tocopherols as a group are not as effective as BHA and BHT.

Commercially prepared tocopherols for use as an antioxidant are most often sold as 50% solutions of tocopherols in vegetable oil and presented for purchase as either Vitamin E T-50 or Tocopherol T-50. Usage rate varies, but common use in bulk oils is between 0.04%–0.5%. Tocopherols are soluble in oil.

Rosemary Oleoresin Extract also known as **ROE** is most likely the favorite antioxidant in the soapmaking industry. It is a thick, viscous, slightly herbaceous solvent extract of the rosemary plant. The carnosal and carnosic acid, found in higher concentration in the oleoresin than in the steam distilled essential oil, are the constituents that give ROE its antioxidant potency. When purchasing, make sure the concentration of carnosic acid is 7% or higher. ROE has been proven to be more effective in many applications and as effective in most applications as BHT. Usage rate is between 0.02%–0.05% by weight. ROE is soluble in oil. Due to its viscous nature, it is important to dilute it with a small amount of the oil you will be incorporating it in to ensure even distribution.

Enhance Antioxidants with Chelating Agents: Trace metals may be found in most of the soapmaking raw ingredients including water, fats, oils and hydroxides. Chelating agents surround and encapsulate the metal ions preventing them from scavenging the fatty acids. They also help the efficiency of the soap by preventing the metal ions from attaching to the hydroxide and forming insoluble soap (aka soap scum). Side benefits to the chelating action are improved clarity in the soap and a boost to the preservative system.

EDTA also known as Ethylenediamine Tetraacetic Acid is a polyamino carbolic acid. Because of its poor solubility in water, EDTA is most commonly used in the form of its various salts, Disodium EDTA, Trisodium EDTA and Tetrasodium EDTA. Tetrasodium EDTA has come under attack recently due to its poor biodegradability. Usage rate is around 0.2% of the salt by weight and 0.1% of the acid by weight.

Citric Acid and Sodium or Potassium Citrate: Citric acid is a weak organic acid commercially manufactured by the bacterial fermentation of sugar. Sodium and potassium citrate are the result of the reaction of citric acid with sodium or potassium hydroxide.

Citric acid can be purchased as anhydrous (containing no water) and as monohydrate (containing one molecule of water). The difference being the monohydrate will weigh more than the anhydrous. Various grades are available including USP, food and lab grade. Food grade citric acid is anhydrous (containing no water) and is 99% pure. Lab grade citric acid is a monohydrate and is generally less pure. Soapmakers can use either. For small quantities, food grade citric acid can be purchased locally in hobby and grocery stores.

Citric acid is often used to neutralize excess alkali in liquid soaps. The citric acid combines with the excess potassium hydroxide forming potassium citrate which serves as a chelating agent as well as a neutralizer of the excess potassium hydroxide. If the soap is neutralized by other methods, sodium citrate can be added with less disruption to the pH, but should be added before neutralization.

Usage rate of citric acid varies to the degree necessary for neutralization and its purity. It is soluble in water and generally made up into a 20% solution for neutralization purposes.

Citric acid and sodium citrate are not as effective as EDTA, but are highly biodegradable and allowed in most certified organic programs.

Usage rate for sodium citrate is between 0.1%–1%.

Gluconic Acid and Gluconates: Gluconic acid is a simple oxidized form of glucose or glucose containing material produced by

fermentation. It is found naturally in royal jelly, honey and wine. Sodium gluconate is the sodium salt of gluconic acid. They have found increasing interest as sequestering and chelating agents due to the overuse of EDTA and the search for more natural alternatives. Gluconates are considered as effective as EDTA and surpass EDTA in alkaline solutions. They are non-irritating to skin and eyes and have humectant properties. Like citric acid, they are effective as pH regulators. Gluconic acid, although much weaker, can be used to neutralize excess potassium hydroxide in liquid soap production. They are stable at boiling point. Gluconates are highly biodegradable and allowed in most organic certified programs. Sodium gluconate used at a rate of 0.4% by weight is considered an effective level as a chelating agent. Usage rate for gluconic acid is generally 0.75–1.00%.

Microbial Rancidity

Microbial rancidity is caused by microorganisms such as bacteria, mold and yeast using their enzymes to break down the chemical structure of the fat or oil. The result is the typical rancid odor, color and destabilization of the product. Water needs to be present in order for this to occur.

Controlling microbial growth begins with purchasing oils and fats from reputable suppliers. They assure the moisture content of the oil or fat is within acceptable parameters to prevent microbial growth.

Proper storage of the oils away from high humidity and heat is also important. Sweating of the container could potentially increase the moisture content of the oil or fat.

Antioxidants and chelating agents prevent oxidative rancidity. Microbial rancidity requires preservation of the oil from moisture during production and storage as well as the use of bactericides in products containing both water and oil.

Consider the Use of a Preservative: One of the most often debated issues regarding liquid soap is whether or not it requires a preservative. It would appear the answer could easily be found through challenge testing, however it is not as simple as it seems. Challenge testing would most certainly answer the question for the specific formula presented and for certain bacteria, but not for all formulas or all bacteria. Below, an incomplete argument ensues based on information gathered. Hopefully, enough of the dots have been connected to lead the soapmaker to an informed decision regarding the use of preservatives in liquid soaps.

Some argue the high pH of liquid soap prevents the growth of bacteria, fungus and mold as they only grow in or near neutral pH between 6 and 8.5. The reality is bacteria, mold, and yeast are capable of growing in a pH of 1 to 11! While most bacteria have a difficult time growing in overly acidic or alkaline environments, there are bacteria that have the ability to adapt to the changes through self-correction or self-protection; these bacteria become alkali or acid tolerant. There are also various bacteria that thrive in extreme pH levels. Types of bacteria are categorized into three groups.

- Acidophile – bacteria which grow in a pH between 1–5.9.

- Neutralphile – bacteria which grow in a pH between 6–9.

- Alkaliphile – bacteria which grow in a pH between 9–11.

Alkaliphile is the one that grabs the attention of the soapmaker. So, what is it? Is it a remote bacteria found only deep within the ocean or alkaline lakes where the likelihood of contamination of a product is so remote as to be nonexistent, or is it more familiar to the environment around us? It is both. Alkaliphilic bacteria can be found in the natural environment in places such as Octopus Spring in Yellowstone Park, Mono Lake in California as well as every day garden soil. It can also be found in artificial environments including industrial waste water effleurage, sewage plants, feces and incrustations on urinals.

While most of the pathogenic (disease causing) bacteria are found within the neutralphile group of bacteria (pH between 6–9), there are known pathogenic alkaliphilic bacteria including *Alcaligenes faecalis* and *Vibrio cholera*.

Alcaligenes faecalis is found in the soil and water and is of main concern to the poultry industry. It is generally considered non-pathogenic to humans but on rare occasions it has been known to cause disease to immune compromised patients through contact with infected medical devices.

Vibrio cholera causes cholera in humans. It lives naturally on plankton in shallow brackish water and is contracted through the consumption of contaminated food or water. It is spread by infected human waste. The best defense against cholera outbreaks, interestingly enough, is thorough washing with soap and water! In experiments carried out in the late nineteenth and early twentieth century, solutions of 4–5% soap of any description were found to completely kill all cholera bacteria within ten minutes. It is still the best defense for preventing widespread contagion. The 2010 cholera epidemic in Haiti was exacerbated in large part due to the shortage of soap and clean water.

While the chances of becoming infected by pathogenic bacteria in liquid soap is extremely remote, it is important to understand non-pathogenic bacteria can contaminate the soap and lead to oxidation, rancidity and instability.

Another argument found in discussion groups is the water in liquid soap is bound by the action of the soap and therefore impervious to microbiologic contamination.

While it is true most products containing water are subject to bacterial contamination there are exceptions. The exceptions are based upon a phenomenon called water activity. So the question is, does water activity play a role in liquid soap?

Water Activity

Water activity is the amount of free water available to enter and leave a product as opposed to water that is chemically bound in the product. Only the "free" water is available for microbial contamination. The example most often cited is honey. The main ingredients in honey are sugar and water, which would appear highly susceptible to microbiological growth. However, the water in the honey is bound by the sugar and therefore makes it unavailable to microorganisms. The same can be said for bar soap. Approximately 15% of water is left in a final bar of soap. Enough of the water is chemically bound by the hydrogen ions and glycerin left in the soap to make preservatives unnecessary. Can the same be said for liquid soap?

First it's important to understand how water activity is measured in a product. The value of water activity is written as a_w. The a_w value of pure water is 1 while a product containing absolutely no water has an a_w of 0.00.

Water activity measurement identifies what type of organisms can grow at specific a_w values. The following chart will be helpful in decoding water activity test results into useful information.

Type of Microorganism	A_w required to prevent growth
Most Pathogenic Bacteria	≤ 0.91
All Pathogenic Bacteria	≤ 0.85
Most Molds and Yeasts	≤ 0.80
All Microorganisms	≤ 0.60

Published results of cosmetics previously tested by Decagon Devices, Inc. which will be of interest to the soapmaker are found in the following chart.

Cosmetic	A_w of Product
Soap, Creamed	0.567
Soap, Bar Regular	0.740-0.757
Soap, Bar with Glycerin	0.659-0.759
Soap, Bar with Glycerin and Lanolin	0.856

Decagon Devices, Inc. graciously agreed to test four samples of liquid soap.

The samples were all from the same batch of liquid soap paste:

- Sample one was the undiluted soap paste.

- Sample two was diluted to approximately 35% soap with no additives.

- Sample three was diluted with water and glycerin to a 40% soap solution then emulsified and thickened with hydroxyethylcellulose.

- Sample four was diluted to a 35% soap solution. Glycerin was added at 12.5% of the soap solution and salt was added at 2% of the soap solution for viscosity.

It is important to mention additives such as salt, glycerin and cellulose derivatives reduce the a_w of a formula, thus the attempt to bind the water as much as possible in the above samples. The chart below will show just how well they fared.

Sample Soap	A_w of Sample
1. Soap Paste, Undiluted	0.866
2. Liquid Soap 35%	0.984
3. Liquid Soap 40% w/Glycerin & Cellosize	0.938
4. Liquid Soap 35% w/Salt & Glycerin	0.941

In examining the results, the undiluted soap was the only sample which came close to being within a water activity range exempting it from bacterial growth. If a portion of the lye water had been replaced by glycerin, it is likely the a_w of the undiluted soap paste would have been lowered below the 0.85 level in which all pathogenic bacteria will not grow.

That being stated, the results are indisputable liquid soap is a water-based product. As such and by definition, it is vulnerable to bacterial contamination.

In summary, the following conclusions may be drawn:

1. Liquid soap is a water based product and is prone to bacteria, but not necessarily pathogenic bacteria.
2. The possibility of alkaliphile pathogenic bacteria contamination harmful to humans is remote.
3. The possibility of contamination by alkali tolerant neutraphile bacteria, including pathogenic bacteria due to a pH shift is increased if the product is any of the following:

- Superfatted
- Over-neutralized
- Contains protein

If the soapmaker is making neutral soap without additives, the greatest need for a preservative is to extend the shelf life and stability of the product and is a matter of personal choice. However, if for example, the soapmaker is making superfatted goat milk liquid soap, the need for a preservative to prevent contamination by pathogenic bacteria should be required.

If the soapmaker concludes a preservative is necessary, the next question is which preservative? There is a vast array on the market. To compound matters, most of the information about specific preservatives comes from the supplier or an anti-everything-but-mother's milk.com website making it almost impossible to make an informed decision. Breaking the problem down into smaller pieces will help eliminate some of the confusion. The first and most important step is determining what is required of the preservative. Looking at the properties of liquid soap as well as the demands placed upon preservatives from the general public and the regulating bodies of differing countries, the following requirements should eliminate much of the confusion. A preservative for liquid soap should:

- Be a broad spectrum preservative blend effective against bacterial and fungal growth.

- Be stable at a minimum pH of 10.

- Not become inactive with the raw materials in the soap or the packaging.

- Be dermatologically tested and considered non irritating at accepted usage levels.

- Be accepted by the general public as safe.

- Be approved for use globally at least at specified levels of concentration.

- Be affordable.

Common preservatives are listed below by category. Only the preservatives which meet most of the criteria stated above are included.

Formaldehyde Releasers

Formaldehyde releasers are chemical compounds that provide preservative protection of a product through the slow decomposition of the compound thereby releasing formaldehyde into the product. While formaldehyde releasers do not accumulate in the environment, formaldehyde is a known allergen, skin sensitizer and carcinogen. In most cases they are recognized as safe as a cosmetic ingredient in the present practice of use. There is no proof the use of formaldehyde releasers used in cosmetics pose significant exposure to formaldehyde to be a concern. Use is restricted by countries in varying degrees depending upon the specific compound. Most of the formaldehyde releasing preservatives are effective within a pH range of 3–9 which is at or below the pH of most liquid soap. Sodium Hydroxymethylglycinate (SHMG) and Quaternium-15 stand apart from the rest with a pH range of 3.5–10 or 3.5–12 depending on manufacturer.

Sodium Hydroxymethylglycinate (SHMG) is manufactured from glycine and formaldehyde in sodium hydroxide and is available as a 50% aqueous solution. It is:

- Marketed by the trade names

    * Suttocide® A (ISP/Sutton)

    * Nipaguard® SMG (Clariant)

    * Mackstat®SHG (McIntyre Group)

- Most active against bacteria and mold.

- Allowed in the USA as used and in Europe and Brazil at usage levels up to 0.5% maximum concentration.

- Not allowed in Japan.

- Reactive with citral, a chemical found in citrus and other essential oils and will effect a color change in the product ranging from light pink to dark red.

- Effective over a pH range of 3.5–12.

- Stored below 80°F (27°C).

- Used at a varying rate dependent upon manufacturer.

- Added in the cool down phase at a temperature of 140°F (60°C) or lower.

- Added to surfactants before the addition of any salt to the formula.

Quaternium-15, also called Chlorallyl methenamine chloride, is a quarternary ammonium salt used as a preservative. It is sold as a pure powder by Dow Chemical Company under the trade name Dowicil® 200 for use in cosmetics and personal care formulas. It is:

- Considered a broad spectrum preservative strongest against gram-negative bacteria and weakest against fungi.

- Allowed in the USA as used and in Europe and Brazil at usage levels of up to 0.2% maximum concentration.

- Not allowed in Japan.

- Reactive with citral, a chemical found in citrus and other essential oils and may cause formulations to turn yellow.

- Effective over a pH range of 4–10.

- Heat sensitive and should not be used at temperatures above 122°F (50°C).

- Water soluble and may be dissolved in water before incorporating into product.

- Used at a rate of 0.02%–0.2% of total weight of ingredients in finished product.

Isothiazolinone

Isothiazolinones got off to a bad start in the 1970s when the first products to be preserved with them created a rash of dermatological irritations and allergies among its consumers. It was discovered formulators had been advised to use isothiazolinones at the same rate as parabens and formaldehyde releasing preservatives. Scientists soon discovered their error and the correct usage rate was established at 100 times less than the usage rate of most current preservatives on the market! Continuing dermatological concerns are based most often on methylchloroisothiazolinone and its combination with methylisothiazolinones, but all isothiazolinones are prone to irritation and reaction when used above acceptable limits. Usage rate is extremely low with as little as one drop of pure methylisothiazolinone needed to effectively provide adequate preservative activity for a fifty-five gallon barrel of water.

Methylisothiazolinone appears to be the best candidate of the isothiazolinone group for liquid soap preservation with a pH limit of up to 12. It is:

- Considered active against bacteria but weak against fungus.

- Globally accepted with usage restrictions of less than 100 ppm in leave on and rinse off products. Japan restricts the use in products that come in contact with mucous membranes.

- Compatible with a wide range of raw materials and is non-discoloring.

- Heat sensitive and should be added at temperatures below 113°F (45°C).

- Biodegradable and does not bio-accumulate.

- Marketed by the trade names:
    - Neolone® 950 manufactured by Rohm and Haas, Inc.
    - Microcare® MT manufactured by Thor Specialties (UK) Ltd.
    - Optiphen® MIT manufactured by ISP.

- Sold as a 9.5% active ingredient in water with a usage rate of 0.05%–0.1% (48–95 ppm of active ingredients) dependent upon manufacturer.

Phenoxyethanol

Phenoxyethanol is an organic chemical compound of ethylene oxide and phenol. It is widely used as a substitute for parabens. It is:

- A weak biocide and always used in conjunction with other preservatives.

- Considered safe as a cosmetic ingredient in the present practice of use and concentration and is allowed in Europe, Brazil and Japan at a use of up to 1% without additional restrictions.

- Necessary to add the highest recommended dosage to the water in anionic surfactants (soap) or it will act as a nutrient for bacteria.

- Inactivated by highly ethoxylated compounds (polysorbates as an example).

- Heat stable and can be used in formulations up to 185°F (85°C).

- Effective over a pH range of 3–10.

- Sold in different purity grades and usage rate is dependent upon manufacturer.

Natural Preservatives

The life cycle of plants and animals requires efficient decomposition once expired. Understanding this natural evolution reveals the difficulty if not impossibility of satisfactorily preserving cosmetics through the addition of natural ingredients, in safe proportions, at a reasonable cost and without compromise.

There are however, a few ingredients of natural origin which work synergistically with other preservatives of interest to the soapmaker. Of special note are the higher molecular glycols with specific emphasis on caprylyl glycol.

Caprylyl glycol may be synthetic or plant derived. It has humectant properties and is often used in conjunction with other preservatives. Preservative blends containing caprylyl glycol that meet the high pH requirements of liquid soap include:

- **Microkill® PCC** manufactured by Arch Personal Care is a broad spectrum preservative system.
  - **INCI** Phenoxyethanol & Chloroxylenol & Caprylyl Glycol.
  - Contains no parabens, formaldehyde releasers or isothiazolinones.
  - Globally accepted.
  - Has a pH range of 3–10.
  - Is insoluble in water, soluble in oil, mostly soluble in glycerin and soluble in polysorbate 20.
  - Heat stable up to 140°F (60°C).
  - Usage rate is 0.5%–1%.

- **Lincoserve™ HpH – 2** is manufactured by Lincoln Fine Ingredients and is a globally approved broad spectrum preservative system designed especially for use in high pH formulations.
  - **INCI** Benzyl Alcohol & Phenoxyethanol & Caprylyl Glycol.
  - Globally accepted.
  - A clear liquid slightly soluble in water.
  - Has a pH range of 3–10.
  - Formaldehyde, paraben and isothiazolinone free.
  - Usage rate is 0.5%–1.5%.

CHAPTER

SPECIALTY SOAPS

"From the angle of the soapmaker, liquid soaps are in most instances identical products to liquid shampoos. They consist of water solutions of potash soap, usually a coconut oil soap."
~ Modern Soapmaking (THOMMSEN & KEMP, 1937)

Specialty soaps are designed with a specific purpose in mind. The same basic recipes in the foregoing chapters can be made into specialty soaps by the use of additives or dilution. Face and hand soaps are basically recipes further diluted so they rinse easily. Soaps for babies as well as people with sensitive skin require a mild formula, such as the soft oil soap recipes listed in both the paste and gel methods. Medicinal soap formulas can include essential oils that have antibacterial, fungicidal and healing properties and dog soaps can include essential oils known for their insecticidal qualities; however, placing these claims on the label or on your website can have repercussions.

Making claims other than the soap cleanses places the product out of the category of soap and into the cosmetic, drug or pesticide category, depending upon the intended use. As a consumer commodity soap is regulated by the Consumer Product Safety Commission and the labeling is regulated by laws under the Federal Trade Commission as long as the intended use is for cleaning purposes only. If claims are made the soap moisturizes, beautifies, adds shine or deodorizes, it is a cosmetic and is regulated by the FDA and requires specific labeling. If claims are made the soap treats dandruff, poison ivy, eczema, cellulite, promotes sleep or grows hair, it is considered a drug and is regulated by the FDA and an approval process must be followed before it can be sold. If claims are made it has insecticidal properties, it is considered a pesticide and is regulated by the Environmental Protection Agency. The bottom line is once a soapmaker steps out of the Consumer Product Safety Commission umbrella, the rules and regulations become more stringent. If selling one's soap, the soapmaker has the responsibility to understand and comply with any and all regulations pertaining to his/her product.

Two of the most often requested recipes are for shampoo and bubble bath. Both are problematic for the handcrafted soapmaker and mostly for the same reason. Yet there are differences in what is expected from each application with some variations so they will be treated separately.

Shampoos

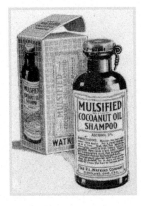

Before the advent of surfactant shampoos, collecting and using rainwater to shampoo the hair was common practice.

Shampoos should be designed to clean the hair and scalp, rinse easily and leave the hair shiny and manageable. Detergent style surfactants can be formulated to achieve all of these goals in one easy step. Liquid soap shampoos will require an extra step with careful formulating.

Synthetic surfactant shampoos can be formulated to be slightly acidic. Cuticle cells in hair shrink and harden when exposed to acidic solutions which equates to smooth and shiny hair. In an alkaline solution such as liquid soap, the cuticles swell and soften which equates to dull, fly away hair. In order to shrink and harden the cuticles after shampooing with liquid soap, or any soap for that matter, it is necessary to add an acidic rinse to the shampooing routine. Vinegar, lemon juice, gluconic acid or citric acid work well to tighten and restore the hair follicle as well as rinse away any soap scum that may have redeposited on the hair. These two steps are necessary to a soap shampoo routine to maintain healthy, shiny and manageable hair.

Although lather is not a bench mark of how well a shampoo cleans, it is a high priority to most consumers. Coconut oil lathers quickly and freely, but the lather isn't as lasting as the shampooing action itself. Hard water also plays a big role in the lathering ability of a soap shampoo. The minerals in hard water form insoluble soap that cuts the lather and detergency power of the soap. Formulating a soap shampoo that will produce long lasting fluffy lather with good cleansing ability is important, but just as important is the addition of the acidic rinse that will aid in removing hard water insoluble soap deposits and restoring the hair to an acceptable pH level. Vinegar is most effective at removing hard water soaps over other acids and it evaporates as it dries, leaving no deposits on the hair. However, many find the odor unpleasant and opt to use other acids. A list of commonly used acid rinses for the hair follows with general usage guidelines. Length, thickness, condition of the hair and the hardness of the water will determine the usage rate. More is not better in the sense too low a pH will risk harming the hair. Maintain the same dilution rate, but increase the quantity if one cup isn't sufficient. Caution is advised acidic rinses may affect colored or permed hair.

Acid Rinses

Citric Acid: ⅛ teaspoon in 1 cup water

Lemon Juice: 1 teaspoon in 1 cup water

Gluconic Acid (purchased as a 40% solution): ½ teaspoon in 1 cup water

Vinegar (purchased as a 5% solution): 1 Tablespoon in 1 cup water

Soap Shampoo Recipes

Many of the recipes already mentioned will make excellent soap shampoos. Just because coconut oil takes the headline on lather, soap gels with their more persistent, although smaller bubbles, do an excellent job as well. Butters and waxes are ideal in stabilizing coconut oils lather and adding conditioning to the hair. Castor oil is cleansing and adds moisture and solubility to the formula. These recipes are designed for the No Paste Method, directions can be found in Chapter 5. The Paste Method can also be used by reserving the dilution water until after the paste has cooked. Directions for the Paste Method can be found in Chapter 3. The glycerin and potassium carbonate in the formula are a bonus that can easily be incorporated into the paste recipe. The glycerin will give moisture and wetting ability to the hair and add clarity to the formula and can be added at any step in the process. The potassium carbonate will help chelate hard water and allow the shampoo to work more efficiently. Add the potassium carbonate to the lye solution before incorporating into the oils if using the paste method.

Coconut Oil, Shea Butter and Castor Oil Shampoo

Expect a much longer cook time due to the shea butter in the formula.

Oils:

29.5 oz. (836 g) Coconut Oil

2.9 oz. (82 g) Shea Butter

3.6 oz. (102 g) Castor Oil

Lye Solution:

7.2 oz. (204 g) Potassium Hydroxide

1.5 oz. (42 g) Sodium Hydroxide

10.4 oz. (295 g) Water

12 oz. (340 g) Glycerin

Dilution:

44.7 oz. (1,267 g) Water

1.8 oz. (51 g) Potassium Carbonate

Total Weight of 40% Soap Solution = 113.6 oz. (3,221 g)

COCONUT OIL, MACADAMIA NUT OIL, MANGO BUTTER AND CASTOR OIL SHAMPOO

Oils:

23.4 oz. (663 g) Coconut Oil

5.4 oz. (153 g) Macadamia Nut Oil

3.6 oz. (102 g) Mango Butter

3.6 oz. (102 g) Castor Oil

Lye Solution:

6.9 oz. (196 g) Potassium Hydroxide

1.5 oz. (43 g) Sodium Hydroxide

10.2 oz. (289 g) Water

12 oz. (340 g) Glycerin

Dilution:

44.4 oz. (1,259 g) Water

1.8 oz. (51 g) Potassium Carbonate

Total Weight of 40% Soap Solution = 112.8 oz. (3,198 g)

COCONUT OIL, JOJOBA OIL AND CASTOR OIL SHAMPOO

Oils:

28.8 oz. (817 g) Coconut Oil

3.6 oz. (102 g) Jojoba Oil

3.6 oz. (102 g) Castor Oil

Lye Solution:

6.9 oz. (196) Potassium Hydroxide

1.5 oz. (43 g) Sodium Hydroxide

10.2 oz. (289 g) Water

12 oz. (340 g) Glycerin

Dilution:

44.6 oz. (1,264 g) Water

1.8 oz. (51 g) Potassium Carbonate

Total Weight of 40% Soap Solution = 113 oz. (3,204 g)

COCONUT OIL, SWEET ALMOND OIL, SHEA BUTTER AND CASTOR OIL SHAMPOO

Oils:

27 oz. (765 g) Coconut Oil

5.4 oz. (153 g) Sweet Almond Oil

1.8 oz. (51 g) Shea Butter

1.8 oz. (51 g) Castor Oil

Lye Solution:

7.1 oz. (201 g) Potassium Hydroxide

1.5 oz. (43 g) Sodium Hydroxide

10.3 oz. (292 g) Water

12 oz. (340 g) Glycerin

Dilution:

44.6 oz. (1,264 g) Water

1.8 oz. (51.03 g) Potassium Carbonate

Total Weight of 40% Soap Solution = 113.3 oz. (3,212 g)

COCONUT OIL, OLIVE OIL, COCOA BUTTER AND CASTOR OIL SHAMPOO

Oils:

25.2 oz. (714) Coconut Oil

5 oz. (142) Olive Oil

2.9 oz. (82 g) Cocoa Butter

2.9 oz. (82 g) Castor Oil

Lye Solution:

7 oz. (199 g) Potassium Hydroxide

1.5 oz. (43 g) Sodium Hydroxide

10.3 (292) oz. Water

12 oz (340 g) Glycerin

Dilution:

44.5 oz. (1,262 g) Water

1.8 oz. (51 g) Potassium Carbonate

Total Weight of 40% Soap Solution = 113.1 oz. (3,206 g)

Bubble Baths

One of the first recipes I tried when I first started making liquid soap was the "Essence of Rosin" bubble bath recipe in *Making Natural Liquid Soaps* (Failor, 2000). While the bubbles were not quite as generous as the bubbles generated from a synthetic surfactant they were more than adequate and the natural essential oils along with the feel of the bath were so much more luxurious. I enjoyed it so much we included it into the Gaily Rebecca Soaps product line, expecting it to be a resounding success. What I did not understand at the time was the effect of the hard water on the product. Unless the customer got their water from a natural spring like we did, or had a water conditioning system, "Gaily Rebecca's Natural Foaming Bubble Bath" was a scum producing, bathtub ring forming, dud of a product.

Trying to build an all-in-one foaming bubble bath with water softening agents in the product is like trying to fight an invading army with a pea shooter. Even pre-softening the bath water with prepackaged water softeners or handcrafted bath bombs is challenging. Giving it a try at my hard water home in Texas, I mixed one cup baking soda with ¼ cup citric acid in the bath. I then added liquid soap one quarter cup at a time, for a total of one cup, to the bath water with the personal sprayer going full blast. No mountain or even hill of bubbles developed, just enough to cover the surface of the water. Until a solution comes along giving better results, the surfactants win in the bubble bath category.

Home Cleaners

Laundry soap and dishwashing soap can be as simple as using the liquid soap you have on hand and adding water softening, detergent boosting, stain fighting ingredients such as OxiClean®, Borateem®, or Arm & Hammer® washing soda according to package directions.

Coconut oil, palm oil, tallow and lard all have excellent cleaning capabilities and recipes including some saturated fat can help clean as well as stabilize the lather when formulating laundry or hand dishwashing soap. Building the laundry soap with more hard water fighting power by adding borax, potassium carbonate or other chelating agents in the dilution phase of the soap will help the soap survive the hard water attack until the laundry booster enters the cycle. General usage rate is ⅛–¼ cup per load depending on type of appliance and size of load. A few samples of "built" laundry or dish soap follow:

COCONUT OIL AND LARD* LAUNDRY AND DISH SOAP

Palm oil can be substituted for the lard without recalculating the lye portion of the recipe.

Oils:

25.2 oz. (714 g) Coconut Oil

10.8 oz. (306 g) Lard or Palm Oil

Lye Solution:

7.1 oz. (201 g) Potassium Hydroxide

1.5 oz. (43 g) Sodium Hydroxide

10.3 oz. (292 g) Water

12 oz. (340 g) Glycerin

Dilution*:

44.6 oz. (1,264 g) Water

14 oz. (397 g) Potassium Carbonate

6 oz. (170 g) Citric Acid

Total Weigh of 40% Soap Solution = 131.5 oz. (3,728 g)

Add the potassium carbonate to the water and stir until mostly dissolved before adding the citric acid. Add the citric acid slowly; it will fizz like a bath bomb and create an exothermic reaction, not as strong as a hydroxide but the water will heat up until the potassium and citric neutralize each other. Bring the water to a boil and proceed as usual.

COCONUT OIL WITH STEARIC ACID LAUNDRY DETERGENT

Oils:

32.4 oz. (919 g) Coconut Oil

3.6 oz. (102 g) Stearic Acid

Lye Solution:

7.5 oz. (213 g) Potassium Hydroxide

1.6 oz. (45 g) Sodium Hydroxide

10.5 oz. (298 g) Water

12 oz. (340 g) Glycerin

Dilution*:

45.1 oz. (1,279 g) Water

7 oz. (198 g) Borax

8 oz. (227 g) Potassium Carbonate

3 oz. (85 g) Citric Acid

Total Weight of 40% Soap Solution = 130.7 oz. (3,705 g)

*Add the potassium carbonate to the water and stir until mostly dissolved before adding the citric acid. Add the citric acid slowly; it will fizz like a bath bomb and create an exothermic reaction, not as strong as a hydroxide but the water will heat up until the potassium and citric neutralize each other. Add the Borax, bring the water to a boil and proceed as usual.

Hard Surface Cleaners

Some handcrafted soapmakers may still remember the original Murphy Oil Soap™ in a jar. Oil soaps were the homemakers' soap of choice when it came to cleaning floors and woodwork and still the better choice over the synthetic surfactants on the market today. Soaps made from coconut oil and saturated fatty acids have a tendency to leave a film when used on hard surfaces and sodium salts are more damaging to surfaces. Soft oils and potassium hydroxide are gentler on countertops and floors. Keep the percentage of coconut oil low and use chelating agents in the dilution phase of the formula to help fight hard water spots. The soaps are made using the Gel Method. Directions can be found in Chapter 8. The paste method can also be used, reserving the dilution phase until after the paste has completed the cook. Directions for use should include a clean rinse to avoid soap build up. General usage rate is 2 oz. (57 g) to a gallon of water for floors and woodwork.

HIGH OLEIC SAFFLOWER OIL SOAP w/ COCONUT OIL

Oils:

32.4 oz. (919 g) High Oleic Safflower Oil
3.6 oz. (102 g) Coconut Oil

Lye Solution:

7.7 oz. (218 g) Potassium Hydroxide
9.8 oz. (278 g) Water
12 oz. (340 g) Glycerin

Dilution*:

43.7 oz. (1,239 g) Water
14 oz. (397 g) Potassium Carbonate
6 oz. (170 g) Citric Acid
Total Weight of 40% Soap Solution = 129.2 oz. (3663 g)

Add the potassium carbonate to the water and stir until mostly dissolved before adding the citric acid. Add the citric acid slowly; it will fizz like a bath bomb and create an exothermic reaction, not as strong as a hydroxide but the water will heat up until the potassium and citric neutralize each other. Add the Borax, bring the water to a boil and proceed as usual.

FLAXSEED (LINSEED) OIL SOAP w/COCONUT OIL

Because linseed oil is considered a drying oil (an oil that hardens to a tough solid film after exposure to air), linseed oil was the favorite soapmaking oil for cleaning hardwood floors and woodwork.

32.4 oz. (919 g) Flaxseed Oil
3.6 oz. (102 g) Coconut Oil

Lye Solution:

7.7 oz. (218 g) Potassium Hydroxide
9.8 oz. (278 g) Water
12 oz. (340 g) Glycerin

Dilution*:

    43.7 oz. (1,239 g) Water

    7 oz. (199 g) Borax

    8 oz. (227 g) Potassium Carbonate

    3 oz. (85 g) Citric Acid

    Total Weight of 40% Soap Solution = 127 oz. (3,606 g)

*Add the potassium carbonate to the water and stir until mostly dissolved before adding the citric acid. Add the citric acid slowly; it will fizz like a bath bomb and create an exothermic reaction, not as strong as a hydroxide but the water will heat up until the potassium and citric neutralize each other. Add the Borax, bring the water to a boil and proceed as usual.

CHAPTER

15

HOW TO FORMULATE LIQUID SOAP RECIPES

"There is nothing more important to the science of soapmaking than a thorough understanding of the fatty acids. Their molecular structures hold the key to understanding why they are fatty, why they are acidic, and why their sodium and potassium salts are soapy."

~ SCIENTIFIC SOAPMAKING (DUNN, 2010)

A soapmaker approached me one day to ask about some problems she was having with her cold process soap. When asked for her formula, she replied it "was a trade secret." When I explained I couldn't help her unless I could examine the oils and lye calculations, but promised not to divulge the information, she reluctantly gave it to me. I'm about to divulge the trade secret: 33% Palm Oil, 33% Coconut Oil and 33% Olive Oil with a 5% lye discount. As everyone who has made cold process soap knows, this is a very basic recipe and will never go down in history as being unique . . . . but she owned it! Furthermore, she had built a very successful business based on one simple formula! Most soapmakers strive to find the perfect recipe that sets them apart from the rest of the crowd.

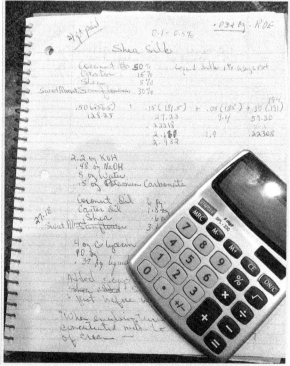

While this book could be filled with soapmaking recipes of every conceivable combination of oils, lyes and additives, it is in the nature of the handcrafted soapmaker to develop

177

one's own special blend of ingredients. In other words, they need to own it!

Much has been explained along the way regarding the ingredients and how they work, but it's time to get down to the actual percentages and calculations necessary in formulating your very own recipes.

The single most important thing a soapmaker should know when formulating recipes is how to calculate the amount of potassium and/or sodium hydroxide necessary to saponify the oils in a given formula. This is the part of the process that sends a large majority of soapmakers running to a lye calculator. If math is not your forte, it's the better option to assure consistent and accurate calculations, but before you head over to the nearest website with a lye calculator, it's important to understand all lye calculators are not created equal. Most fail to take into consideration the moisture and impurities found in potassium hydroxide and none are designed to calculate an excess of lye. At the time of this writing, most liquid soapmakers have found good results with the following lye calculators:

> *A soapmaker does not have to be a math wizard in order to calculate formulas. The use of a lye calculator found on many soapmaking websites in combination with the dilution chart found in Appendix B will simplify the process into a few easy steps.*

- **Handcrafted Soap and Cosmetic Guild** (www.soapguild.org/members/) – The advanced calculator in the members' area of the website allows the soapmaker to plug in one's own saponification values. By using the saponification chart located in Appendix C, which includes varying degrees of excess alkali already calculated into the saponification value, the soapmaker can easily formulate a recipe tailored to a specific grade of potassium hydroxide. The calculator also allows the soapmaker to calculate a combination of both sodium and potassium hydroxide into the formula and the values are consistent with the values in this book.

- **Summer Bee Meadow** (www.summerbeemeadow.com) – The Summer Bee Meadow calculator has already calculated a conservative lye excess of 4 – 6%, however the calculator will allow a negative number to be entered for a super-fatting percentage. Some quick math determined that a -4% would get a potassium hydroxide value very close to a 10% lye excess. While Summer Bee Meadow's calculator will also calculate both sodium and potassium hydroxide into the formula the results, for reasons not understood, come out differently than when calculating the same percentages using the method outlined below, however both approaches work. If using the Summer Bee Meadow calculator, expect the sodium hydroxide portion of the recipe to be about 9% greater and the potassium hydroxide portion 9% less than if calculated by the method below.

Saponification Value

In order to calculate the amount of potassium hydroxide required in a recipe, it is necessary to know the saponification (SAP) value of the oil. The saponification value (SAP) is the number of milligrams of potassium hydroxide required to saponify one gram of fat. It is a measure of the average molecular weight (or chain length) of all fatty acids present in a particular oil. All oils have their own saponification value because they are made up of differing combinations of fatty acids. A saponification chart can be found in Appendix C with the range of values of many of the soapmaking oils. Included in the chart are also the average SAP values for potassium and sodium hydroxide as well as values for calculating excess potassium hydroxide in a formula.

Step 1 Converting Milligrams to Grams

In order to begin calculating the amount of potassium hydroxide necessary to completely saponify fats and oils, it is necessary to compare apples to apples and convert all units of measure into grams (that common denominator math sort of thing). Because the weight of the potassium hydroxide is in milligrams and the weight of the fat is in grams, it is necessary to convert the potassium hydroxide into grams. How many milligrams are in one gram? The answer is 1,000. To convert the amount of potassium hydroxide into grams, divide the saponification value by 1,000. For example: Assuming it requires 189 milligrams of potassium hydroxide to fully saponify one gram of olive oil, you would divide 189 by 1,000 (189 ÷ 1,000 = 0.189).

Step 2 Calculating the Amount of Potassium Hydroxide

Most liquid soaps are formulated with a blend of oils rather than only one. It is necessary to know the percentage of each oil in the formula. An example could be a liquid soap with

60% Coconut Oil   30% Olive Oil   10% Castor Oil.

The average saponification value; already converted into grams for the oil are:

Coconut = 0.2565   Olive = 0.1890   Castor = 0.1815

The next step is to determine the average SAP value for the oils in the formula. This can be accomplished by multiplying the percentage of the oil with its corresponding SAP value and then adding the numbers together. The formula would look like this:

(60% Coconut × 0.2565 SAP) + (30% Olive × 0.1890 SAP) + (10% Castor × 0.1815 SAP) = SAP Value for Oil Blend.

(0.1539 Coconut SAP) + (0.0567 Olive SAP) + (0.0815 Castor SAP) = 0.22875 SAP Value for Oil Blend.

Whether making enough to fill a pint or a barrel, knowing the SAP value of a blend of oils allows the soapmaker to adjust the recipe up or down by multiplying the SAP value of the blend of oils by the total amount of oils in the formula. Since this recipe is already listed in the book as #7 using 36 ounces (1,021 g) of oils in the formula, let's say the soapmaker wants to make a recipe using 50 ounces (1,418 g) of oil instead. The soapmaker would then multiply the amount of oil in the formula:

(50 oz. or 1,418 g) × 0.22875 KOH SAP Value for Oil Blend = 11.4 oz. (323 g) KOH

This would be the amount of potassium hydroxide required to saponify the blend of oils in the recipe; but wait!—this SAP value does not take into account the absorption of moisture or impurities within the potassium hydroxide. An excess of potassium hydroxide is necessary to overcome these impurities as well as to ensure an excess of potassium hydroxide is left in the formula to achieve a truly neutral soap through the neutralization process. The recipes in this book were calculated with 10% excess of potassium hydroxide. To calculate this formula the same requires taking the 11.4 oz. (323.2 g) of potassium hydroxide and multiplying by 110% or 1.10:

11.4 oz. × 1.10 = 12.5 oz. Potassium Hydroxide or (323 g × 1.10 = 356 g Potassium Hydroxide).

By multiplying the percentage of the oils with the total amount of oils in the formula the recipe would then read:

60% Coconut Oil × 50 oz. (1,418 g) = 30 oz. (851 g) Coconut Oil

30% Olive Oil × 50 oz. (1,418 g) =
15 oz. (425 g) Olive Oil

10% Castor Oil × 50 oz. (1,418 g) =
5 oz. (142 g) Castor Oil

0.22875 SAP × 50 oz. (1,418 g) × 1.10 =
12.5 oz. (323 g) Potassium Hydroxide

But what if the soapmaker wanted to use a percentage of sodium hydroxide in the formula? Old time soapmakers used sodium hydroxide to stiffen soft soap in the summer months to prevent it from becoming too thin. It was generally packaged in firkins (small wooden barrels with staves) not fancy glass or plastic leak proof containers as they are now; you can imagine the mess. In the winter months it had a tendency to cloud the soap and become too stiff so it was omitted.

Step 3 Calculating a Recipe with Both Potassium and Sodium Hydroxide

The recipes in the no paste method of soap are calculated with 75% potassium hydroxide and 25% sodium hydroxide. This is the ratio I found that worked well for me, but usage rate can range anywhere from 16%–33% sodium hydroxide to 84%–67% potassium hydroxide. Using the 75% potassium hydroxide and 25% sodium hydroxide ratio of alkali, the next step is to determine the amount of sodium hydroxide required to saponify the 50 oz. (1,417.5 g) of oils and then multiply by 25% to arrive at the amount of sodium hydroxide required to saponify 25% of the oil blend.

By definition, saponification values are based on the molecular weight of potassium hydroxide. Because potassium hydroxide is heavier than sodium hydroxide it is necessary to convert the saponification value to reflect the weight of the sodium hydroxide. Knowing the molecular weight of potassium hydroxide is 56.1 and the molecular weight of sodium hydroxide is 40 the next step is to create a conversion ratio by dividing 40 by 56.1 and arrive at a constant conversion number of 0.713 (40 ÷ 56.1 = 0.713). This is the key to convert any saponification values into sodium hydroxide values. Taking the saponification value of 0.22875 and multiplying it by the constant conversion number 0.713 will give the sodium hydroxide value of the blend of oils.

0.22875 SAP KOH × 0.713 Constant = 0.1631 SAP NaOH

Calculating 25% Sodium Hydroxide Solution:

$$0.1631 \text{ SAP NaOH} \times 50 \text{ oz. } (1{,}418 \text{ g}) =$$
$$8.15 \text{ oz. } (231 \text{ g}) \text{ NaOH} \times 25\% = 2 \text{ oz. } (58 \text{ g}) \text{ NaOH}$$

Calculating 75% Potassium Hydroxide Solution with 10% Excess Alkali:

$$0.22875 \text{ SAP KOH} \times 50 \text{ oz. } (1{,}418 \text{ g}) \text{ oil} =$$
$$11.43 \text{ oz. } (324 \text{ g}) \times 75\% = 8.57 \text{ } (243 \text{ g}) \times 110\% =$$

$$9.4 \text{ oz. } (267 \text{ g}) \text{ KOH}$$

The recipe now reads:

Oils:

30 oz. (851 g) Coconut Oil

15 oz. (425 g) Olive Oil

5 oz. (142 g) Castor Oil

Lye Solution:

9.4 oz. (267 g) Potassium Hydroxide

2 oz. (58 g) Sodium Hydroxide

Now for the water portion of the recipe:

Step 4 Calculating the Water

There must be enough water in a soapmaking formula to sufficiently dissolve the hydroxides and carry the lye solution to all parts of the pan, but not so much water the concentration of the lye solution is too weak to work effectively. All recipes in this book are first calculated with 66.7% anhydrous soap and 33.3% water. Anhydrous soap is simply the combined total of the oils and dry potassium and/or sodium hydroxide in the formula. The dilution chart in Appendix B can be used to determine the amount of water necessary or the soapmaker can make the calculations. To calculate the amount of water necessary, first add the oils and the hydroxides. Using the recipe in the example above, there is a total of:

50 oz. (1,418 g) of Oils + 9.4 oz. (267 g) KOH + 2 oz. (58 g) NaOH = 61.4 oz. (1,743 g) Anhydrous Soap.

The next step is to determine the total amount of water and oils in a 66.7% anhydrous soap solution. It is necessary to divide the amount of anhydrous soap by 66.7% or 0.667

> 61.4 oz. (1,743 g) Anhydrous Soap ÷ 0.667 (66.7%) = 92.05 oz. (2,613 g) Soap Paste.

> Oils + Alkali = Anhydrous Soap. Anhydrous Soap × 0.5 = Amount of Water.

Subtract the total amount of the soap paste from the amount of anhydrous soap to determine the amount of water in the formula.

> 92.05 oz. (2,613 g) Soap Paste–61.4 oz. (1,743 g) Anhydrous Soap = 30.65 oz. (870 g) Water.

> Or

Using the chart in Appendix B, multiply the anhydrous soap by 0.5:

> 61.4 oz. (1,743 g) Anhydrous Soap × 0.5 = 30.7 oz. (870 g) Additional Water to make a 66.7% Soap Paste

The recipe now reads:

Oils:

30 oz. (851 g) Coconut Oil

15 oz. (425 g) Olive Oil

5 oz. (142 g) Castor Oil

Lye Solution:

9.4 oz. (267 g) Potassium Hydroxide

2 oz. (58 g) Sodium Hydroxide

30.65 oz. (870 g) Water

Total Soap Paste:

92.05 oz. (2,610 g) Soap Paste

Step 5 Calculating a No Paste or Gel Soap Recipe

Most recipes can be made into either a No Paste or Gel Soap recipe depending upon the mix of oils in the formula. Formulas containing 40% or more of coconut oil will make a No Paste liquid soap. Formulas containing 70% or more of soft oils will make a gel soap but recipes of 75% or greater are recommended.

All recipes in this book begin with a 66.7% anhydrous soap paste and are further diluted upon trace for the No Paste and Gel Making Method. Most recipes found in *Making Liquid Soap* (Failor, 2000) and on the internet are around 61%–65% anhydrous soap. The dilution water can be adjusted up or down to accommodate the variation. A dilution chart can be found in Appendix B to make the calculations a bit easier.

Calculating the Glycerin. The percent of glycerin in the No Paste and Gel soap recipes are calculated at 33% of the oils in the recipe or roughly 10% of a 35% anhydrous soap solution. Multiplying the amount of oils × 33% is just easier math. The percentage of glycerin along with other ingredients in this process came about by trial and error. A philosophy of "if it ain't broke don't fix it" was adopted whenever a method worked consistently. If the soapmaker decides to change the percentages, be advised that too much glycerin will dampen the lather and cause the soap to feel sticky on the skin.

> Total Oils × 0.33 = Amount of Glycerin

To arrive at the amount of glycerin necessary, multiply the amount of oils × 33%.

$$50 \text{ oz. } (1{,}417 \text{ g}) \text{ Total Oils} \times 0.33 = 16.5 \text{ oz. } (468 \text{ g}) \text{ Glycerin.}$$

The glycerin will be added to the lye portion of the recipe, so it will be necessary to subtract the amount of glycerin from the water portion to maintain a 66% anhydrous soap paste.

$$30.7 \text{ oz. } (870 \text{ g}) \text{ Water} - 16.5 \text{ oz. } (468 \text{ g}) \text{ Glycerin} = 14.2 \text{ oz. } (403 \text{ g}) \text{ Water.}$$

> Total Oils × 0.05 = Amount of Potassium Carbonate

Potassium Carbonate. The percent of potassium carbonate in the No Paste and Gel soap recipes are calculated at 5% of the oils in the formula which is a little less than 1.5% of a 35% anhydrous soap solution. Again this calculation was based on ease of calculations and trial and error. The potassium carbonate will be added to the dilution water.

$$50 \text{ oz. } (1{,}418 \text{ g}) \text{ Total Oils} \times 0.05 = 2.5 \text{ oz. } (71 \text{ g}) \text{ Potassium Carbonate.}$$

Step 6 — Calculating the Dilution Water:

The dilution water is calculated at 60% water to 40% anhydrous soap. It is important to use the least amount of water necessary to keep the soap fluid. Concentrated soap cooks much faster than diluted soap.

To calculate the dilution water in a recipe is as simple as 1) adding water equal to the amount of the anhydrous soap in the formula, 2) using the dilution chart in Appendix B or 3) dividing the amount of anhydrous soap by 40% and then subtracting the total weight of the soap solution by the weight of the soap paste.

$$61.4 \text{ oz. } (1{,}743 \text{ g}) \text{ Anhydrous Soap} \div 0.40 =$$
$$153.5 \text{ oz. } (4{,}352 \text{ g}) - 92.05 \text{ oz. } (2{,}608 \text{ g}) \text{ Soap Paste} = 61.4 \text{ oz. } (1{,}743 \text{ g}) \text{ Dilution Water.}$$

> When calculating the recipes for this book, it took me several recipes to realize the dilution water for a 40% soap stock was equal to the anhydrous soap in the formula. Boy, did that make my job so much easier!

The final recipe with the Glycerin, Potassium Carbonate and Dilution Water Calculated

Oils:

30 oz. (851 g) Coconut Oil

15 oz. (425 g) Olive Oil

5 oz. (142 g) Castor Oil

Lye Solution:

9.4 oz. (267 g) Potassium Hydroxide

2 oz. (58 g) Sodium Hydroxide

14.2 oz. (403 g) Water

16.5 oz. (468 g) Glycerin

Dilution Water

61.4 oz. (1,743 g) Dilution Water

2.5 oz. (71 g) Potassium Carbonate

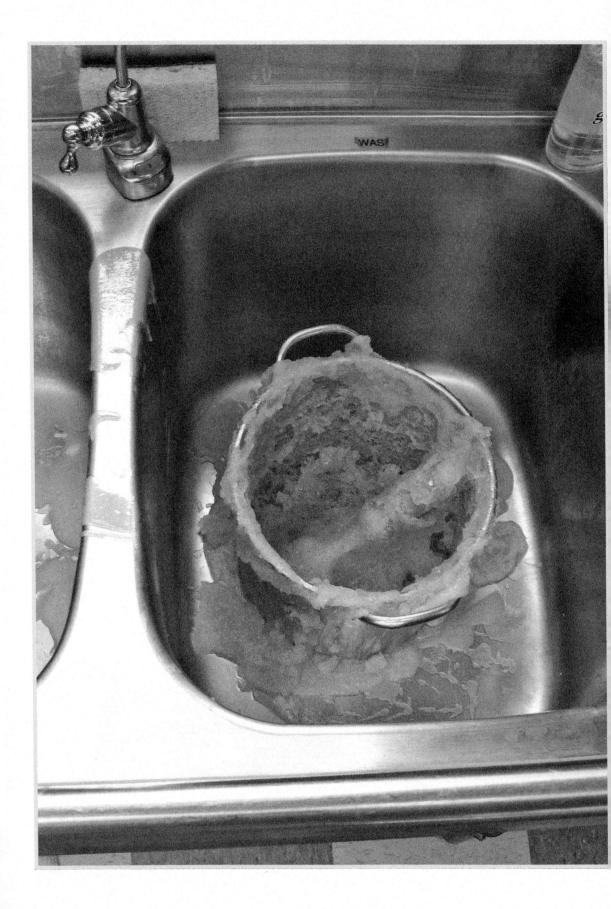

CHAPTER

TROUBLESHOOTING PROBLEMS

"I have come by the following questions quite naturally—first by asking them myself, and then by having them asked of me by others.... Some are scientific or technical. Some have clear answers, and some are controversial."

∼ THE SOAPMAKER'S COMPANION (CAVITCH, 1997)

Soap ingredients separate: The soap started tracing and separated, looking more like applesauce instead of paste. No amount of stirring will get it to come back together, even with the addition of water. There are curds on top and a clear dark liquid on the bottom.

Solution 1 Most likely there is too much alkali in the formula. Try adding Castor oil in 2 ounce (57 g) increments to the formula to bring it into emulsion. Castor oil is the quickest of the soapmaking oils to saponify.

Solution 2 Don't confuse separation in a No Paste recipe during the cook with separation of the oils and lye leading to trace. In No Paste, the soap will often separate into clear saponified soap on the bottom with a creamy blend of partially saponified soap and alkali on the top during the cook. This is just a matter of periodically stirring the emulsion back together until all of the soap has saponified to a clear broth.

> Separation may occur if the raw soap is cooked at too high a temperature. Add a couple of ounces of cold water to the soap solution and stick blend to try and bring back into an emulsion. Turn down the heat and proceed as usual once the formula comes back together.

Soap paste doesn't reach Vaseline stage:

Solution 1 Perform a phenolphthalein test to determine if there is enough alkali in the formula. If the paste tests clear, more alkali is required. Dilute to a 40% soap concentration and follow the directions for neutralizing unsaponified oils beginning on page 42.

Solution 2 If the soap is still alkaline, check the temperature of the soap. If the center of the paste is below

180°F (82.2°C), turn up the heat and cook it for another hour or two and test it again. Some soaps require up to 12 hours to fully cook. Cook time is largely dependant upon the size of the batch and the type of oils in the formula.

Gloppy goop floating on top of soap:

Solution This is most likely caused by too high a concentration of soap in the solution. Increase the rate of dilution. The thicker the glop, the more water should be added.

Milky film floating on top of soap:

Solution This is most likely caused by unsaponified or superfatted oils in the formula. Try cooking the soap longer with an additional 1 oz. of a 50% lye solution.

Soap is cloudy after dilution:

Solution 1 Hard water may be the culprit. Use only soft or distilled water to dilute the soap stock.

Solution 2 If the soap tested clear and clouded after neutralization it could be over neutralization of the soap stock. The best thing to do at this point is to sequester the soap and allow the free fatty acids to settle to the bottom.

Solution 3 Soaps high in saturated fatty acids made using the No Paste method will be crystal clear while warm, but cloud upon cooling. There may be too high a percentage of saturated fatty acids in the formula. Refrigerate the soap stock for 48 hours and allow the soap to return to room temperature. This should precipitate the clouding saturated fatty acid soap to the bottom of the container and the clear soap can be carefully poured into a separate container.

Soap clouds in cold weather:

Solution This is normal. The potassium carbonate in the gel and no paste method drastically reduce the cloud point in low temperatures, but all liquid soap will cloud. The soap will clear once the temperature rises.

Soap remains cloudy: The soap has sequestered for two weeks with no change in the clarity of the soap.

>   *Solution*   The soap stock is most likely too highly concentrated. Highly concentrated soap holds the insoluble soaps and free fatty acids in suspension. Dilute the soap further and allow it to sequester.

White sludge on bottom of container of soap:

>   *Solution*   Depending on when the sludge developed, it is most likely:
>
>   1. A matter of superfatting with insufficient sequestering if it develops within a few weeks of manufacture.
>   2. Taking already diluted soap and diluting further for foamy pump bottles. Dilution will a) lower the pH of the product and b) allow unsaponifiables to more readily precipitate out of solution and cause sedimentation.
>   3. Caused by oxidation and bacterial contamination if it develops later during its shelf life.

Soap clouds after adding fragrance: The soap was perfectly clear until the essential/fragrance oils were added. Now it's cloudy.

>   *Solution*   Some fragrances and essential oils will cloud the soap due to the solvents, waxes or acidic nature of the EO or fragrance oil and most will always cloud when added to a cooled soap. If the soap was cool when the fragrance was added, allow a few days for the soap to clear. If the soap still remains cloudy, warm up the soap stock and add perfumers' alcohol at a rate of 50%–100% of the weight of the essential/fragrance oil. This usually clears it right up.

Soap separated or thickened after adding fragrance:

>   *Solution*   This is most likely due to oil-soluble fragrance oils. The addition of polysorbate 20 or lecithin will bring the fragrance oils back into an emulsion with the water based liquid soap. Start with 50% of the weight of the fragrance oil and stir gently, but thoroughly to incorporate. Up to 200% of the weight of the fragrance oil can be added to the formula.

APPENDIX

BATCH CODE SHEET

Name			Date		
Name of Soap			Batch Code		

Oils, Fats or Waxes	Weight	Supplier	Lot Number	Expiration Date	Comments

Water and/or Liquids	Weight	Source/Supplier	Lot Number	Expiration Date	Comments

Base	Weight	Source/Supplier	Comments
KOH			
NaOH			
K_2CO_3			

Method of Cook	Cook Time & Temperature	Clarity	Phenolpthalein Test	Comments

Dilution Liquid	Weight	Source/Supplier	Lot Number	Expiration Date	Comments

Additional Ingredients	Weight	Source/Supplier	Expiration Date	Comments

Sequester Date	Sequester Time	Appearance	Odor	Comments

APPENDIX

DILUTION CHART FOR FORMULATING LIQUID SOAP

Desired Concentration of Soap	Starting @ 100% Soap Multiply by	Starting @ 66% Soap Multiply by	Starting @ 50% Soap Multiply by	Starting @ 40% Soap Multiply by	Starting @ 35% Soap Multiply by	Starting @ 30% Soap Multiply by	Starting @ 25% Soap Multiply by	Starting @ 20% Soap Multiply by
66%	0.5	—	—	—	—	—	—	—
50%	1.0	0.33	—	—	—	—	—	—
40%	1.5	0.67	0.25	—	—	—	—	—
35%	1.86	0.91	0.43	0.14	—	—	—	—
30%	2.33	1.22	0.67	0.33	0.17	—	—	—
25%	3.0	1.67	1.0	0.6	0.4	0.2	—	—
20%	4.0	2.33	1.5	1.0	0.75	0.5	0.25	—
15%	5.66	3.45	2.33	1.67	1.33	1.00	0.67	0.33

APPENDIX C

SAPONIFICATION VALUES OF COMMON SOAPMAKING OILS

Oil	SAP Value	KOH Average Value @ 0%	NaOH Average Value @ 0%	KOH Alkali Value @ 8%	KOH Alkali Value @ 10%	KOH Alkali Value @ 12%
Sweet Almond	188–200	.1940	.1370	.2095	.2134	.2174
Apricot Kernel	185–199	.1920	.1370	.2074	.2112	.2150
Avocado Oil	177–198	.1875	.1337	.2025	.2063	.2100
Babassu	245–256	.2505	.1786	.2705	.2756	.2805
Canola Oil	182–193	.1875	.1337	.2025	.2063	.2100
Castor	176–187	.1815	.1294	.1960	.1990	.2030
Cocoa Butter	192–200	.1960	.1397	.2117	.2156	.2195
Coconut	248–265	.2565	.1828	.277	.2820	.2870
Hempseed	190–195	.1925	.1373	.2079	.2118	.2156
Jojoba	88–96	.0920	.0656	.0994	.1012	.1030
Macadamia Nut	190–200	.1950	.1350	.2106	.2145	.2184
Mango Butter	183–198	.1905	.1358	.2057	.2096	.2134
Olive	183–195	.1890	.1347	.2041	.2079	.2117
Palm	190–209	.1995	.1422	.2150	.2190	.2230
Palm Kernel	244–250	.2470	.1760	.2668	.2717	.2766
Rice Bran	181–189	.1850	.1319	.1998	.2035	.2072
High Oleic Safflower	180–198	.1890	.1347	.2041	.2079	.2117
Shea Nut Butter	178–198	.1880	.1340	.2030	.2068	.2106
High Oleic Sunflower	187–194	.1905	.1358	.2057	.2096	.2134

APPENDIX D

PERCENT OF MOST COMMON FATTY ACIDS IN SOAPMAKING OILS

Oil	Stearic 18:0	Palmitic 16:0	Oleic 18:1	Linoleic 18:2	Linolenic 18:3	Lauric 12:0	Myristic 14:0	Ricenoleic 18:1–
Sweet Almond	1–10	4–13	43–70	20–34				
Apricot Kernel	0.5–1.2	4.6–6	58–65.7	28.5–33	0.5–1.0			
Avocado	0.4–1	9–18	56–74	10–17	0–2			
Babassu	1.8–7.4	5.2–11	9–20	1.4–6.6		40–55	11–27	
Canola	1.1–2.5	3.3–6	52–67	16–25	6–14		0–0.2	
Castor	0.9–2	1–2	2.9–6	3–5	0–0.5			88
Cocoa Butter	31–37	25–27	31–35	2.8–4.0	0.1		0.1	
Coconut	2.3–4.8	7.7–10.2	5.4–9.9	0.56–2.1	0–0.2	45.1–50.92	16.8–21.09	
Corn	0–3.3	9.2–16.5	20–42.2	44.7–65.6	0.5–1.5	0–0.3	0–0.3	
Cottonseed	2.1–3.3	18.1–26.4	14.2–21.7	46.7–58.3	0–0.4	0–0.2	0.6–1.0	
Emu	8–11	19–25	41–54	9–22	0.2–1.5		0.3–0.6	
High Oleic Safflower	1.5–2.0	5–6	74–80	13–18	0–0.2		0–0.1	
High Oleic Sunflower	3–5	3–5.2	70–92	2–20	trace			
Lard	5–24	20–32	35–62	3–16	0–0.5		0.5–2.5	
Macadamia Nut	2–4	8–9	56–59	2–3			0.6	
Mango Butter	31–48.88	4–12	38–50	3–6.34	0.7–1.0	0.3–0.4	0.11–0.5	
Olive	0.5–5	7.5–20	55–83	9–21	0–1.5		0–0.1	
Palm	3.5–6.5	40–48	36–44	6.5–12	0–0.5		1.8	
Palm Kernel	1.3–3	6.5–10.3	12–21	1–3.5	0–0.7	40–55	14–18	
Peanut	1.9–4.4	8.3–14	36.4–67.1	14–43	0.0–0.1	0–0.1	0–0.1	
Rice Bran	2–4	16–28	38–48	16–36	0.2–2.2		0.5–0.7	
SoyBean	3.0–5.4	9.7–13.3	17.7–28.5	49.8–57.1	5.5–9.5	0–0.1	0–0.2	
Soy Bean High Saturate	17.5	21.9	9.4	37.5	11		0.1	
Shea Butter	36–41	4–8	45–50	4–8	0–0.4	0.4	0.3	
Sunflower	2.5–7	5–8	13–40	48–74	0–0.3	0–0.1	0–0.2	
Safflower	1.9–2.9	5.3–8	8.4–30	67.8–83.2	0–0.1		0–0.2	
Tallow	25–40	20–37	31–50	1–5			1–6	

APPENDIX

ESSENTIAL OILS AT A GLANCE

Essential Oil	Flashpoint	Specific Gravity	Maximum Concentrate @ 1% Category 9	Maximum Concentrate @ 5% Category 9	IFRA Specifications, EU Allergens
Bergamot	108–136°F (42.22–58.33°C)	0.860–0.875 @ 72°F (22.22°C)	100% Rinse Off	100% Rinse Off	IFRA – yes EU – yes
Cassia	194°F (90°C)	1.04–1.07 @ 20°C	5%	1%	IFRA – no EU – yes
Atlas Cedarwood	212°F (100°C)	0.925–0.95 @ 25°C	100%	100%	None
Clary Sage	200°F (93.9°C)	0.886–0.929 @ 25°C	100%	40%	IFRA – yes EU – yes
Clove Bud	239°F (115°C)	1.038–1.06 @ 25°C	55%	11%	IFRA – no EU – yes
Eucalyptus	120.2°F (49°C)	0.90–0.93 @ 25°C	100%	100%	IFRA – no EU – yes
Lemon Eucalyptus	165°F (74°C)	0.86–0.87 @ 20°C	100%	100%	IFRA – yes EU – yes
Egyptian Rose Geranium	204.8°F (96°C)	1.016	100%	100%	IFRA – yes EU – yes
Geranium	172.4–179.6°F (78°–82°C)	0.883–0.890 @ 25°C	100%	100%	IFRA – yes EU – yes
Guaiacwood	282°F (138.89°C)	0.96–0.975 @ 25°C	100%	100%	IFRA – no EU – no
Lavandin	144°F (62°C)	0.88–0.903 @ 25°C	100%	100%	IFRA – yes EU – yes
Lavender	149°F (65°C)	0.875–0.888 @ 25°C	100%	100%	IFRA – yes EU – yes
Lemongrass	163°F (73°C)	0.869–0.894 @ 25°C	100%	100%	IFRA – no EU – yes
Oakmoss	190°F (87.8°C)	1.02–1.035 @ 25°C	10%	2%	IFRA – yes EU – yes
Palmarosa	212°F (100°C)	0.880–0.894 @ 25°C	100%	100%	IFRA – no EU – yes
Patchouli	190°F (88°C)	0.95–0.975 @ 25°C	100%	100%	IFRA – no EU – yes
Peppermint	170.6°F (77°C)	0.9–0.91 @ 25°C	100%	40%	IFRA – no EU – yes
Peru Balsam	> 200°F (> 93.33°C)	1.095–1.11 @ 25°C	40%	8%	IFRA – no EU – yes
Rosemary	104°F (40°C)	.894–.912 @ 25°C	100%	100%	IFRA – no EU – yes

ESSENTIAL OILS AT A GLANCE (CONTINUED)

Essential Oil	Flashpoint	Specific Gravity	Maximum Concentrate @ 1% Category 9	Maximum Concentrate @ 5% Category 9	IFRA Specifications, EU Allergens
Dalmatian Sage	124°F (51.1C)	0.903–0.925 @ 25°C	100%	100%	IFRA – no EU – yes
Spearmint	145°F (62.78°C)	0.917–0.934 @ 25°C	100%	40%	IFRA – yes EU – yes
Sweet Orange Folded	115°F (46°C)	0.868–0.878 @ 25°C	100%	100%	IFRA – yes EU – yes
Tea Tree	136°F (57.8°C)	0.885–0.906 @ 20°C	100%	100%	IFRA – no EU – yes
Thyme	132°F (56°C)	0.915–0.93 @ 25°C	100%	40%	IFRA – no EU – yes
Ylang Ylang	180°F (82°C)	0.908–0.930 @ 25°C	100%	100%	IFRA – no EU – yes

GLOSSARY

Acid: Any substance that produces hydrogen ions when combined with water. Acid solutions have a pH lower than 7 and react with alkalis to form salts. Examples of how acids are used in soapmaking include the combination of fatty acids with potassium or sodium hydroxide to produce soap (a salt of the fatty acid). Acids such as citric acid are used as a reactant to neutralize over-alkaline liquid soap solutions by forming potassium citrate (the potassium salt of citric acid).

Additives: Any ingredient not necessary to the soapmaking process. Examples of additives include fragrance, color, preservatives and extracts.

Alkali: Any water soluble substance that produces hydroxide ions when combined with water. Alkali solutions have a pH greater than 7. Examples of how alkalis are used in soapmaking include the combination of sodium hydroxide with fatty acids to form bar soap and the combination of potassium hydroxide with fatty acids to form liquid soap. Alkalis are also referred to as bases.

Alternative Liquids: Any liquid used to replace all or a portion of the water during the soapmaking process. Common alternative liquids used in soapmaking include milk, aloe vera, specially denatured alcohol, glycerin, wine and beer.

Anhydrous Soap: Soap containing no water. Anhydrous soap in a soapmaking recipe is calculated by adding the total amount of oil in the recipe with the total amount of dry sodium and/or potassium hydroxide in the recipe.

Antioxidants: Compounds known to increase the resistance of fats and oils to oxidation and rancidity, thus extending the shelf life of the product by maintaining the physical integrity of the molecules. Examples of antioxidants commonly used by soapmakers include Vitamin E T-50 and Rosemary Oleoresin Extract (ROE).

Batch Code Sheet: A written summary documenting the process used in manufacturing each batch of a product. The batch code sheet should include a code number identifying the specific batch, a list of the raw ingredients including the suppliers and any lot or serial numbers, the name of the soapmaker, the date of manufacture and any pertinent observations made during the process.

Butters: A group of natural fats including saturated as well as unsaturated fatty acids which are solid or semi-solid at room temperature. They are generally very stable and contain a wealth of unsaponifiables beneficial to skin and hair. Some butters on the market today, such as coffee butter, are artificially produced using hydrogenated oils and extracts.

Caustic: Any substance with the ability to burn or corrode something by chemical action. Sodium and potassium hydroxide are also referred to as caustic soda and caustic potash.

Centigram Balance: A weighing scale that reads to hundredths of a gram (0.01 gram).

Chelating Agents: Ingredients incorporated into soap and detergents to prevent metal ions from attaching to the hydroxide by either trapping or suspending the minerals out of solution and preventing them from scavenging the fatty acids, thus reducing the amount of soap scum produced by use in hard water. Examples of chelating agents commonly used in soapmaking include sodium and potassium citrates, gluconates, sorbates and silicates. Chelating agents will not protect a product from oxidation or microbial contamination.

Consumer Product Safety Commission (CPSC): An independent agency of the U.S. government whose purpose is to regulate the sale and manufacture of all products in the U.S. not under the jurisdiction of other federal agencies such as the Bureau of Alcohol Tobacco and Firearms and the Food and Drug Administration. Soap is regulated under the CPSC unless claims are made that would classify it as a cosmetic or drug.

Contamination: The unintended introduction of undesirable physical, biological or chemical elements making a product unfit for use. Physical contaminants include foreign objects such as hair or glass. Biological contaminants include organisms such as flies, bacteria or viruses. Chemical contaminants include substances such as residual cleaning solutions or pesticides.

D&C Dyes: Colorants approved for use in drugs and cosmetics following regulations set by the FDA.

Denatured Alcohol: Ethyl alcohol, also called grain alcohol or ethanol) with a small percentage of other ingredients added to make it undrinkable and therefore non-taxable. Various ingredients are used as a denaturant depending on the purpose of use. It is sold as 190 or 200 proof (95% ethanol and 100% ethanol respectively) before any denaturant is added. Not all types of denatured alcohols are appropriate for use in soap and not all are skin safe.

Distilled Water: Water in which much of the impurities have been removed through steam distillation. The purest form of water readily available to soapmakers.

Emulsion: The suspension of a liquid in another liquid that normally cannot be combined by slowly adding one ingredient to another with vigorous agitation. This suspends tiny droplets of one liquid through another. Combining oil and lye with vigorous stirring creates an emulsion.

Extracts: A liquid or powdered substance containing the active ingredient of a plant in concentrated form.

Exothermic Reaction: A chemical or physical reaction that releases energy in the form of heat. When potassium or sodium hydroxide is dissolved in water the chemical reaction of the water and alkali substantially increase the temperature of the solution; it can get very hot.

FDA: An acronym for the United States Food and Drug Administration. The FDA is an agency within the U.S. Department of Health and Human Services responsible for overseeing the safety and effectiveness of food, drugs, cosmetics, tobacco products and medical devices.

FD&C Dyes: Colorants approved for use in food, drugs and cosmetics following regulations set by the FDA.

Fatty Acids: Any of a group of organic acids derived from the breakdown of animal or vegetable fats, oils and waxes. Fatty acids of interest to the soapmaker are palmitic, stearic, myristic, lauric, oleic and ricinoleic acids. Linoleic and linolenic fatty acids are usually avoided. When chemically reacted with an alkali they form soap.

Free Alkali: Any remaining alkali left unreacted by the acids in a formula.

Free Fatty Acids: Fatty acids not attached to other molecules. Free fatty acids in liquid soap occur as a result of over neutralization or the degradation of the fats or oils in the formula.

Glycerin: A byproduct of saponification but can also be produced synthetically from propylene, a petroleum product. It is a sweet, viscous sugar alcohol with humectant and solvent properties. In liquid soapmaking glycerin provides clarity, speeds saponification and allows for a more concentrated soap solution.

Hard Water: Water containing a high content of minerals. Hard water is formed when the water flows through limestone, chalk or iron ore deposits before making its way to the surface. Using hard water in liquid soap formulations will cause the soap to be cloudy due to the formation of insoluble soaps within the formula (soap scum).

Humectant: A hygroscopic substance that absorbs or attracts moisture from the air and aids other substances in retaining moisture. Glycerin is an example of a humectant used in liquid soapmaking.

Hydrolysis: The decomposition of a chemical compound by its reaction with water. Saponification is the alkaline hydrolysis of fatty acid esters.

Hygroscopic: A substance that attracts moisture from the air. Potassium hydroxide is hygroscopic.

Hydrophobic: A substance that repels water. Waxes are hydrophobic.

Hydroxyethylcellulose (HEC): A thickening agent synthetically derived from cellulose, which may be used to thicken liquid soap.

International Fragrance Association (IFRA): A global membership organization for the fragrance industry promoting and supporting the safety, sustainability and integrity of fragrance products by establishing and supporting standards of safe usage levels to be carried forward by their membership. IFRA standards are accepted by the US FDA and by regulatory agencies of the EU and other countries around the world.

Isopropyl Alcohol: Also referred to as "rubbing alcohol", is synthetically derived from propylene, a petroleum product. It is generally diluted with water most commonly as a 70% isopropyl solution but can be found or ordered at most drugstores as a 91% solution. 91% isopropyl alcohol may be substituted for denatured alcohol for test purposes.

Infusion: A method of extracting chemical constituents from a plant by steeping in water, oil or alcohol. Similar to making a cup of tea from a tea bag.

Lipids: A group of organic compounds including fats, oils, waxes, oil soluble vitamins (sterols) and other related compounds.

Lye: Alkali such as potassium hydroxide or sodium hydroxide in an aqueous solution.

Lye Discount: Using a smaller percentage of alkali than required to completely saponify all the oils in a given formula, resulting in free fats not being made into soap.

Material Safety Data Sheet (MSDS): A document outlining any potential danger of a product with guidelines on how to handle the product in a safe manner as well as what to do in the case of any spill or accident.

Neutral Soap: A soap containing no excess oils or lye despite the fact it is still alkaline.

Neutralize: To render harmless usually by applying its opposite force. Soapmakers neutralize excess alkali in a formula with citric acid to ensure the soap is safe and gentle to the skin.

Oxidation: The degradation of the quality of an oil or fat by the chemical reaction of the oxygen in the air with the oil or fat. Oxidation leads to rancidity. Antioxidants are used to retard oxidation.

pH: The measure of the acidity or alkalinity of a solution on a scale of 0–14 with 7 being neutral. Any solution with a pH less than 7 is considered acidic. Any solution with a pH greater than 7 is considered alkaline. Pure water with a pH of 7 is considered neutral.

Phenolphthalein: A chemical compound used as a pH indicator. It indicates the degree of acidity or basicity of the object being measured by a change in color. It generally appears colorless in solutions below pH 8 and hot pink in the presence of free alkali. It is used in soapmaking as an aid in achieving neutral soap.

Potassium Carbonate: Also referred to as potash or pearl ash with a chemical formula K_2CO_3. It is a weak alkali. It is used in liquid soap formulas as a chelating agent and for its ability to keep a concentrated soap solution fluid.

Potassium Hydroxide: Also referred to as caustic potash or KOH (its chemical formula) is a strong alkali used in the production of liquid soaps due to its ability to absorb water more easily than its counterpart sodium hydroxide. It is a hazardous material and should be handled appropriately.

Preservative: An ingredient used to aid in the prevention of the growth of microorganisms such as bacteria, mold, yeast, fungus and spores in a product.

Rancidity: A term used to describe the off odor, off color and decomposition of oil or fat as the result of oxidation and/or bacterial contamination of the oil or fat.

Saponification: The chemical reaction of a fat or oil with a strong base or alkali (such as potassium hydroxide) to create a salt of the fatty acid, what is commonly referred to as soap and glycerin.

Saponification Value (SAP): The amount of potassium hydroxide required to convert a particular fat or oil into soap.

Saturated Fatty Acid: Fatty acids which have all the hydrogen the carbon can hold and therefore have no double bonds. Because they have no double bonds they resist oxidation, are more stable and have longer shelf lives than unsaturated fatty acids. Because they are so dense they are usually solid at room temperature. Lauric, myristic, palmitic and stearic acids are examples of saturated fatty acids.

Separation: When particles in an emulsion break apart. Possibly forming different layers or distinct masses of material.

Sequester: Allowing a solution to rest for a period of time undisturbed so sediment can settle to the bottom and the clear liquid can be decanted from the top.

Sequestering Agents: (See chelating agents).

Shelf Life: The length of time a product remains effective, free from deterioration and usable. Rancidity and microbial contamination affect the shelf life of liquid soap.

Soap Scum: The saponification of mineral salts in hard water with the alkali in soap to form insoluble salts technically called lime soaps. Lime soaps leave deposits on the surfaces it comes in contact with and is difficult to rinse away. Soap scum is inhibited by chelating agents.

Sodium Chloride: Common table salt.

Sodium Hydroxide: Also referred to as caustic soda or NaOH (its chemical formula). It is a strong alkali used mainly in the production of bar soaps but a small percentage is often used in liquid soap to increase viscosity and boost lather. Sodium hydroxide is a hazardous material and should be handled properly.

Soft Oils: Oils high in unsaturated fatty acids.

Soft Soap: A term used for a liquid soap made predominantly with soft oils.

Soft Water: Water lacking any appreciable amount of dissolved minerals. Water can be naturally soft or hard water can be softened with water treatment systems.

Sorbitol: A sugar alcohol found naturally in various fruits and plants and produced from corn syrup. Sorbitol adds clarity, brilliance and emollience to a liquid soap formula.

Super-Fatting: Using a greater amount of oil than required to completely saponify all the alkali in a given formula, resulting in the presence of extra oils.

Titration: The process of adding one solution to another solution in carefully measured amounts until a definite reaction is achieved and the added volume may be accurately measured. Titration is used to determine how much citric acid solution is required to neutralize excess alkali in liquid soap.

Trace: The point at which an emulsion thickens sufficiently a portion of the solution when dribbled over the surface leaves a trail.

Triglycerides: An ester made up of three molecules of fatty acids attached to one molecule of glycerol.

Unsaponifiables: The hydrocarbons, fatty alcohols and pigments that make up in large part the nutrients of the oil or fat and are unreacted by the saponification process.

Unsaturated Fatty Acids: Fatty acids which have one or more double bonds connecting two carbons to each other. Unsaturated fatty acids are liquid at room temperature. Oleic, ricinoleic, linolenic and linoleic acids are examples of unsaturated fatty acids.

Water Activity: A term used to describe the amount of free water available to enter and leave a product as opposed to water that is chemically bound in the product. Only the free water in a product is available for microbial contamination.

Waxes: Composed of one long-chain fatty acid bonded to a long-chain alcohol group, waxes are similar to fats and can be saponified into soap except they contain no glycerin and are extremely hydrophobic.

BIBLIOGRAPHY

Books

Cavitch, Susan Miller. *The Soapmaker's Companion*. Pownal, VT: Storey Books, 1997.

Cottone Elyse, 2009. *Use of Natural Antioxidants in Dairy and Meat Products: A Review of Sensory and Instrumental Analyses*. Masters Thesis, Kansas State University, Manhattan, KS.

Deite, Carl, Engelhardt, F. Wiltner. *The Soap Maker's Handbook of Materials, Processes and Receipts for Every Description of Soap*. NY: Henry Carey Baird & Co, 1912.

Dunn, Kevin. *Scientific Soapmaking*. Farmville, VA: Clavicula Press, 2010.

Failor, Catherine. *Making Natural Liquid Soaps*. Pownal, VT: Storey Books, 2000.

Firestone, David, Editor. *Physical and Chemical Characteristics of Oils, Fats and Waxes, 2nd Edition*. Washington, D.C.: AOCS Press, 2006.

Gale, Marie. *Soap & Cosmetic Labeling; How to Follow the Rules and Regs Explained in Plain English*, Broadbent, OR: Cinnabar Press, 2008.

Hurst, George H. *Soaps,* London: Scott, Greenwood & Co., 1898.

Hurst, G.H. and Simmons, W.H. *Textiles Soaps and Oils: A Handbook on the Preparation, Properties and Analysis of the Soaps and Oils Used in Textile Manufacturing, Dyeing and Printing*. London, 1914.

Lawless, Julia. *The Illustrated Encyclopedian of Essential Oils. The Complete Guide to the Use of Oils in Aromatherapy and Herbalism*, NY: Barnes and Noble Inc. by arrangement with Element Books Limited, 1999.

Lesser, Milton A. *Modern Chemical Specialties.* NY: MacNair – Dorland Company, 1950.

Poucher, William A., PH.C. *Perfumes, Cosmetics & Soaps, Volume II Being a Treatise on Practical Perfumery.* NY: D. Van Nostrand Company, Inc.,1932.

Schwartz, Leonard. *Sanitary Products.* NY, NY: MacNair-Dorland Company, 1943.

Stanislaus, Stanley, I.V. and Meerbott, P.B. *American Soap Maker's Guide.* New York, NY: Henry Carey Baird & Co., Inc., 1928.

Steinberg, David C. *Preservatives for Cosmetics,* Second Edition. Carol Stream, IL: Allured Publishing, 2006.

Thomssen, E.G. *Soap-Making Manual, A Practical Handbook on the Raw Materials, Their Manipulation, Analysis and Control in the Modern Soap Plant.* NY: D. Van Nostrand Company, 1922.

Thomssen, E.G. and Kemp, C.R. *Moden Soapmaking.* NY, NY: MacNair-Dorland Company. 1937.

Thomssen, E.G., Ph. D., McCutcheon, John W. M.A., F.C.I.C., *Soaps and Detergents.* NY: MacNair-Dorland Co., 1949.

Websites

Bodner Research Web. "Acid Base Pairs, Strength of Acids and Bases and pH." *http://chemed.chem.purdue.edu/genchem/topicreview/bp/ch11/conjugat.php*

Brenntag Specialties, Inc. "Safflower Oil High Oleic Product Data Sheet." *http://www.brenntagspecialties.com/en/downloads/Products/Personal_care/Textron/PDS_SAFFLOWER_OIL_HIGH_OLEIC_TX008285.pdf*

Darwin Chemical Company. "Stearic Acid Triple Pressed Vegetable, Safety Data Sheet." *http://www.darwinchemical.com/Products/Specs/Q270.html*, Stearic Acid, Triple Pressed, Vegetable.

Decagon Laboratories. "Water Activity." *http://www.wateractivity.org/*

Fontana Jr., PhD., Anthony, *Understanding Water Activity for Reduced Microbial Testing Using USP Method,* Pullman, WA: Decagon Devices. *www.aqualab.com*

Gadberry, Rebecca James, "Ingredient Review: The Safety of Paraben Substitutes." *http://www.skininc.com/skinscience/ingredients/17013151.html.*

Huebsch, Russell, "The Effects of pH on Bacterial Growth." *http://www.ehow.com/about_5695904_effects-ph-bacterial-growth.html*.

International Fragrance Association. *http://www.ifraorg.org*

International Jojoba Export Council. "Jojoba Oil Specification Sheet." *http://www.ijec.net/downloads/specifications.pdf*.

Macadamia Processing Company. "Macadamia Oil Specification Sheet." *http://www.intermac.com.au/pdf/spec_oil.pdf*. Macadamia Processing Company.

Potassium Hydroxide. *http://www.potassium-hydroxide.com*

Sunflower Oil High Oleic Product Data Sheet. *http://www.e.com/uploads/resources/51/warner_.pdf*. High Oleic Sunflower.

United Nations Conference on Trade and Development. "International Agreement on Olive Oil and Table Olives, 2005." *http://www.unctad.info/upload/Infocomm/Docs/olive/cxs_033e.pdf*.

Wilhelm, Pape Louis Friedrich Hans, Umbach, Inventor. *Process for the manufacture of nondepositing liquid soaps. U.S. Patent # 2105366*. Publication Date 1/11/1938. *www.google.com/patents/*

RESOURCES

Vendors

A+ Soapmaking Studio—*www.SoapmakingStudio.com*—Soapmaking supplies and chemicals including cosmetic grade hydroxyethylcellulose, potassium carbonate, KOH, NaOH, phenolphthalein, pipette droppers.

AAA Chemical—*www.aaa-chemicals.com*—Soapmaking supplies and chemicals including NaOH, KOH, Gluconic Acid

Brambleberry—*www.brambleberry.com*—General soapmaking supplies including bottles and jars, colorants, extracts, fragrances, oils, additives and liquid soap calculator.

Certified Lye—*www.certified-lye.com*—NaOH, KOH, Potassium Carbonate and Phenolphthalein.

Chemistry Store—*www.chemistrystore.com*—Large selection of chemical ingredients with an emphasis on soapmaking supplies.

Citrus and Allied—*www.citrusandallied.com*—Essential and fragrance oils.

Columbus Food; Soapers Choice—*www.soaperschoice.com*—Soapmaking oils, butters and waxes; bulk, pail and gallon.

Essential Depot—*www.essentialdepot.com*—General soapmaking supplies; oils, butters, essential and fragrance oils, potassium and sodium hydroxide.

Essential Wholesale & Labs—*www.essentialwholesale.com*—Extensive selection of natural ingredients for soap, cosmetics and aromatherapy, including private labeling.

Handcrafted Soap and Cosmetic Guild—*www.soapguild.org*—Membership community of soapmakers offering programs and services including insurance.

Kaufman Container—*www.kaufmancontainer.com*—Wholesale stock bottles by the case.

Lebermuth & Sons—*www.lebermuth.com*—Essential and fragrance oils including natural fragrances.

Liberty Natural—*www.libertynatural.com*—Botanical extracts and essential and fragrance oils including Guaiacwood and Oakmoss.

MyGMPRecords.com—*www.mygmprecords.com*—Subscription based service that helps in keeping all records required by Good Manufacturing Guidelines.

Nashville Wraps—*www.nashvillewraps.com*—Wholesale distributor of gift and gourmet packaging products.

Old Will Knot Scales—*www.oldwillknotscales.com*—Large selection of scales and balances for the soapmaker at good prices.

Paper Mart—*www.papermart.com*—Packaging supplies, gift boxes and more.

Papilio—*www.papilio.com*—Waterproof inkjet blank labels.

Shay and Company, Inc.—*www.shayandcompany.com*—Soapmaking oils, butters, essential and fragrance oils and specialty ingredients.

SKS Bottle & Packaging, Inc.—*www.sks.com*—Wholesale supplier of bottles, jars and closures with small minimum pieces.

Soapalooza—*www.soapalooza.com*—Extensive selection of soap and cosmetic ingredients.

Soapies Supplies—*www.soapies-supplies.com*—General soapmaking supplies including a handy label applicator.

Specialty Bottle—*www.specialtybottle.com*—National supplier of glass, plastic and metal containers and closures.

Summer Bee Meadow—*www.summerbeemeadow.com*—Excellent source of information and lye calculator

Voyageur Soap and Candle Co. Ltd.—*www.voyageursoapandcandle.com*—General soapmaking supplies including glyceryl stearate for pearlizing soap.

Wholesale Supplies Plus—*www.wholesalesuppliesplus.com*—General soapmaking supplies including bottles and jars, colorants, extracts, fragrances, oils and additives.

INDEX

Acid, 8–10, 197
Acid rinses, 166, 167
Acidophile, 155
Additives, 26, 139–147, 197
Alcohol, 35, 36, 75, 110, 123, 189, 198
Alkali, 10–15, 24, 31, 42, 197
Alkali tolerant neutraphile, 158
Alkaliphile, 155, 158
Alternative liquids, 141, 142, 197
Anhydrous soap, 32, 183, 197
Antioxidants, 21, 151, 152, 197
Apricot kernel oil, 18, 45, 193, 194
Atlas cedarwood, 103, 195
Avocado oil, 18, 45, 193, 194
Bacteria, 154–156
Base, 122, 125
Basic, 8, 9
Batch Code Sheet, 4, 6, 191, 197
Bergamot, 101, 102, 195
Blenders, 122
Blending essential oils, 120–123
Botanical extracts, 146, 147
Bubble bath, 170, 171
Butters, 18, 19 198

Canola oil, 18, 45, 193, 194
Cassia, 102, 103, 195
Castor oil, 18, 19, 194
Centigram balance, 35, 37, 198
Chelating agents, 41, 152–154, 198
Citric acid, 35–37, 153, 167
Clary sage, 103, 104, 195
Clove bud, 104, 195
Cocoa butter, 19, 193, 194
Coconut oil, 19, 20, 60, 193, 194
Crockpot, 28
Dalmatian sage, 115, 195
Denatured alcohol, *See* Alcohol
Dilution, 32, 33, 63, 83, 192
Dish soap, 171, 172
Egyptian (rose) geranium, 107, 195
Essential oils, 99–123
EU Cosmetic Directive, Annex III, 7th Amendment, 99
Eucalyptus, 105, 106, 195
Fats, 8, 16
Fatty acid, 8, 16, 17, 199
FD&C dyes, 95, 199
Fixators, 122, 123
Flash point, 121, 195, 196
Formaldehyde releasers, 159, 160

Free Alkali, 30, 31, 39, 40, 199
Free fatty acids, 149, 188, 189, 199
Gel soaps, 79–93
Geranium, 106, 195
Gluconic acid, 153, 154, 167
Glycerin, 21, 60, 200
Glycol distearate, 97
Guaiacwood, 108, 195
Hard surface cleaners, 173
Hydrolysis, 31, 200
Hydrophobic, 18, 200
Hydroxyethylcellulose, 75, 76, 200
Hygroscopic, 11, 200
IFRA standards, 99, 195, 200
Infusions, 97, 146, 147, 200
Isothiazolinone, 161, 163
Lard, 17, 171, 194
Laundry soap, 171
Lauric acid, 16, 194
Lavandin super, 108, 195
Lavender, 108, 109, 195
Lemon eucalyptus, 106, 195
Lemon juice, 167
Lemongrass, 110, 195
Linoleic acid, 18, 150, 194
Linolenic acid, 18, 150, 194
Lye, 11, 201

211

Lye discounts, 139, 201
Macadamia nut oil, 18, 45, 193
Mango butter, 193
Microbial rancidity, 150, 154
Milk soaps, 143
Modifiers, 122, 123, 125
MSDS sheet, 6, 11, 201
Myristic acid, 16, 17, 193
Natural colorants, 96, 97
Natural preservatives, 162
Neutral soap, 9, 10, 201
Neutrality, 30, 42, 63, 83
Neutralizer, 36–43
Neutraphile, 155
No Paste method, 59–64
Oakmoss, 110, 195
Oleic acid, 17, 18, 45, 194
Olive oil, 18, 45, 51, 194
Oven process, 29
Oxidative rancidity, 150–154
Palm oil, 49, 172, 194
Palmarosa, 111, 112, 195
Patchouli, 112, 195
Pathogenic, 155, 156
Pearl soap, 17, 97
Peppermint, 112, 113, 195
Peru balsam, 113, 114, 195

pH, 8, 9, 201
Phenolphthalein, 4, 6, 10, 187, 201
Phenoxyethanol, 162, 163
Potassium carbonate, 13, 60, 61, 80, 81, 171, 184, 201
Potassium hydroxide, 10–12, 24, 42, 43, 179, 180–182, 193, 201
Preservatives, 158–163,
Rancidity, 149–155, 202
Rice bran oil, 18, 45, 194
Ricinoleic acid, 18, 194
RIFM, 99, 100
Rosemary, 114, 115, 195
Rosemary oleoresin extract (ROE), 152
Safety, 1–6, 23, 61, 80
Salt, 73, 74, 202
Saponification, 8, 13, 202
Saponification value, 13–15, 179–182, 193, 202
Saturated fatty acids, 16, 17, 173, 188, 202
Scales, 4, 5
Separation, 27, 202
Sequester, 43, 188, 202
Sequestering Agents, 140, 141
Shampoos, 97, 166–170
Shea butter, 19, 194

Sodium chloride, 73, 74, 202
Sodium hydroxide, 12, 13, 24, 60, 181, 193, 202
Soft oils, 18, 45, 51, 81, 173, 184, 203
Sorbitol, 60, 140, 141
Spearmint, 115, 116, 196
Specially denatured alcohol, *See* Alcohol
Specific gravity, 121, 195
Stearic acid, 17, 40–42, 172, 194
Stick blender, 4, 25
Superfatting, 139, 140, 158, 203
Sweet almond oil, 18, 45, 194
Sweet orange, 116, 117, 196
Sweeteners, 122, 123
Tea tree, 117, 194
Thickening, 73–77
Thyme, 117, 118, 196
Tocopherols, 21, 152
Trace, 26, 203
Triglyceride, 8
Unsaponifiables, 20, 21, 203
Unsaturated fatty acids, 17, 203
Vinegar, 9, 166, 167
Water, 16, 182, 196
Water activity, 156, 157
Water bath, 28
Waxes, 8, 18, 19